How to Ace Calculus:
The Streetwise Guide

How to Ace Calculus: The Streetwise Guide

Colin Adams
Williams College

Joel Hass
University of California, Davis

Abigail Thompson
University of California, Davis

W. H. Freeman and Company
New York

Cover Designer: Patricia McDermond
Text Designer: Victoria Tomaselli

Library of Congress Cataloging-in-Publication Data

Adams, Colin Conrad.
    How to ace calculus : the streetwise guide / Colin Adams, Joel
Hass, Abigail Thompson.
        p.   cm.
    Includes index.
    ISBN 0-7167-3287-4 (textbook). — ISBN 0-7167-3160-6
    1. Calculus—Study and teaching.   I. Hass, Joel.   II. Thompson,
Abigail.   III. Title.
    QA303.3.A33   1998
    515′.071—dc21   97-51944
                                                                    CIP

Printed in the United States of America

First printing 1998

W. H. Freeman and Company
41 Madison Avenue, New York, NY 10010
Houndmills, Basingstoke RG21 6XS, England

**To all the students whose
lifelong ambitions were dashed
on the cliffs of calculus**

CONTENTS

Introduction

If you are reading this introduction then this book is probably not for you. This book is directed at calculus students who have better things to do with their time than read wordy preambles that won't be on the exam. But just in case you haven't actually bought this book yet and are considering a purchase while flipping through the pages in a bookstore, we'll tell you what it's all about.

If you want to know the tricks of the trade that will make learning the material of first-semester calculus a piece of cake, then this is the book for you. If you want to learn lots of cool things while having a good time, then this is the book for you. If you want to carry around a book that makes people think you are surfing the wave of knowledge, then this is the book for you.

Do you remember being in a class and being hopelessly confused? Perhaps your attention wandered at some important moment, or the lecturer thoughtlessly slipped into ancient Greek when explaining the basic idea. After class, you asked your brainy friend over a cup of coffee, "What was going on in that class?" Your friend explained it all in five minutes flat and made it crystal clear. "Oh," you said, "is that all there is to it? Why didn't they just say that in the first place?" Later, you wished that friend was around to explain all the lectures to you.

This book aims to play the role of your friend. It gives informal explanations of the key topics of calculus, getting across the ideas without the technical details and fine print that would be found in a formal text. This book does not substitute for a calculus textbook, but it should make it much easier to figure out what the textbook is talking about.

If you approach it with the right point of view, learning calculus can be not only a mind-expanding experience but also fantastic fun, just about as good as something not involving whipped cream and maraschino cherries can get. This book is going to tell you how calculus is taught, how to get the best teachers, what to study, and what is likely to be on exams. This is the stuff we wish we'd known when we had to take calculus. So, enough stalling. Why don't you go up to that nice cashier, plunk down some money and buy this book, and we can talk more after?

Exactly Who and What Is Your Instructor?

2.1 Choosing an instructor

Here we give a brief introduction to mathematicians, their pecking order, and their identifying characteristics.

READ THIS BEFORE CHOOSING AN INSTRUCTOR.

Understanding mathematicians is a lot like bird watching. You need to know enough distinguishing features to say, "Ah, a yellow-bellied sapsucker" with conviction.

Choosing the best instructor is the single decision most likely to determine whether your calculus experience will be a series of intellectual delights or whether you schedule dental appointments during calculus lectures because they are less painful.

You can often figure out the specifics about your instructors by looking at their doors. Generally a small sign will be posted containing some clue to the instructor's official title. There are several possibilities:

A. Permanent faculty, tenured (sign on door says Professor or Associate Professor). Tenured means that they cannot be fired, even if they are grossly incompetent. Associate Professors are a rung below Professors.

Sometimes this is because they are at an earlier stage in their career, sometimes because their career stalled after they were discovered hiding in the chimney of the dean's apartment.

B. Permanent faculty, untenured (sign on door says Assistant Professor). These people can be fired, but if they are, it will not be for reasons related to their ability to teach calculus. In Europe, Assistant Professors really are assistants, whose job is to mow the Professor's lawn, carry the Professor's briefcase, and teach the Professor's class. In the United States, the Assistant title just means that they are in the preliminary, untenured stage of their career.

C. Visitors (sign on door says Visiting Professor, or Visiting Assistant Professor). "Visiting" means that their welcome is due to expire at the end of one or two years. It does not necessarily mean that they have anywhere to go afterward.

D. Temporary faculty (sign on door says: Lecturer or Instructor or Adjunct Professor). Some colleges hire temporary faculty mainly to teach classes. This may mean that they really care about their teaching.

E. Graduate students (sign on door just gives their name, with no title, or has some pseudo-title like Adjunct Instructor).

F. No sign on door: A very bad omen. It may mean that the instructor is too disorganized to post a sign or that revengeful former students keep ripping it down. Perhaps the instructor is trying to avoid previous generations of students? Investigate further.

G. No door: Danger, danger. Could mean that the instructor is deemed unworthy of an office. This makes it hard to hold office hours. Also could mean that you're looking in the wrong building.

Almost everyone in categories A, B, C, and D has a Ph.D.

Permanent faculty members, with their ranks, should be listed in the course catalog, where you can look them up. At large universities, permanent faculty are usually either the people who do research in really high-level mathematics or the people who have been around forever and control the math department political machine. Sometimes the best teachers are found in this group. The worst teachers are often found here. These faculty will be teaching all the advanced mathematics courses, as well as calculus. Some find teaching calculus a chore, an obligation that must be tolerated in order for them to be able to do their research. Others truly enjoy it.

At many smaller institutions, the faculty aren't expected to do much research. Instead, the emphasis is on teaching, and you will find many faculty members who put a tremendous amount of time and thought into their teaching. This sometimes translates into better teachers. It is also true that professors at such schools often have to teach two to three times as many classes as their counterparts at the research universities. So, although they are not distracted by their attempts to prove that the convex core of a hyperbolic 3-manifold is compact, they are often distracted by the demands of hundreds of students they are teaching in a given semester in four different classes.

Distinguishing between categories C (visitors) and D (temporary faculty) is tricky but can be important. Some visiting faculty members are on sabbatical or on leave from their real home somewhere else. They may be visiting a colleague at your university, or perhaps your campus is located near a good surfing spot. They will return home at the end of the year, blissfully unconcerned with the eviscerating essay you turned in on your student evaluation. Pleasing the students ranks about 1.5 on a scale of 100 for these faculty. On the other hand, you can sometimes stumble across a real gem here, someone who comes from a campus where teaching is job one.

Other visiting faculty may be new faculty who have just finished their Ph.D.s and are currently teaching their first class. They may have some title like "Visiting Assistant Professor" or something with the word "postdoc" in it. Their main concerns are their research and getting another job the following year. They can't totally blow off teaching if they want to get another job, and though they usually try to do a good job, their teaching abilities vary wildly.

Lecturers and temporary instructors are usually hired on a year-to-year basis, primarily to teach. Whether their job is renewed at the end of their contract probably depends heavily on their teaching skills. Their future meals may depend on their student evaluations. Pleasing the students is extremely important to them. If your instructor has been at the university for more than a year and does not have a title containing the word "professor," the instructor is probably temporary. These instructors usually teach only calculus and precalculus classes, and if they have been around for awhile, it's very likely they do a superb job. If not, they may be related to the department chair.

Graduate students can be great instructors, but they usually don't have much experience. Their teaching ability is all over the map, from very high to very low. One problem with graduate students is that it's hard to get anything past them. They remember taking calculus all too well. Some graduate students do have years and years of experience—their minds were addled in the 1960s, and they've been in graduate school ever since. You can usually spot these people by their snarly gray hair, slurred speech, and the ratty ponchos they wear when it rains.

In most cases grad students have little in the way of a teaching track record for you to research. Your best strategy with them is to attend the first class or two, and if it looks disastrous, hightail it out of there.

Often, professional status correlates with amount of office space—if all else fails, you can deduce the status of an instructor by checking out the office. Estimate its square footage and divide by the number of names on the door. Add the floor number and multiply by -1 if there are no windows. The bigger the resulting number, the more important your instructor is in the departmental hierarchy.

There are, of course, many exceptions to these wild generalizations. We've known visiting professors who are dedicated and brilliant teachers, and tenured faculty members who devote their lives to their students. Fortunately, there's a more accurate and quite easy way to find out who the brilliant lecturers are.

Trade Secret   *To find out who the best lecturers are: Ask!*

Everyone around the department knows the good teachers. Try dropping in on a couple of faculty members and ask them. Try the younger ones; they don't know enough to mislead you. The secretaries will know who generates waiting lists and who generates a lot of dropped classes and incompletes. The graduate students will know. If there is a math club, its members will be dying to talk to you about this. (Get an excuse ready so you can leave after 10 minutes. They will be prepared to discuss it all afternoon.)

Some colleges publish student reviews of instructor performance. You (or your parents or the taxpayers) are paying a fortune for four years of college. Shell out a few bucks if necessary and take the time to help make the difference between enlightenment and boredom.

Trade Secret   *To find the best instructor: Go to several classes and stay with the best!*

In large universities you may have a choice of five or ten different classes in the same subject. In smaller places, hope that teaching is taken seriously. By asking around you should be able to find two or three likely candidates who do a great job. Try them out.

Given a choice, why not pick the best instructor? You will usually be able to tell the first day who is going to be good and who will put you to sleep. Set your standards high. Don't make the mistake that many students do of assuming that it is your own fault if the class is incomprehensible.

Trade Secret   *If a class you are prepared for is completely incomprehensible, it is probably the fault of the instructor.*

If more than 10 percent of the class is asleep on the first day, that is a BAD sign. If a clear handout is distributed explaining the class, and if the instructor holds lots of office hours, that is a GOOD sign. If you are unsure whether you are in a French 101 class or a math class, that is a BAD sign. If this book is required in the course, that is a GOOD sign. A VERY GOOD sign.

Colleges do not like students checking out classes like this because it makes more paperwork for them and makes it harder to predict class size. Some college administrators would like you to pay your tuition and stay out of sight. Don't worry about them—get your money's worth.

If you happen to be taking calculus in high school, then a lot of this advice is probably irrelevant. You have no choice. You will be assigned a particular teacher, usually the only one teaching calculus that year, and it doesn't matter how much you kick and scream, that's who you get. You'll just have to cross your fingers and hope for the best.

On the other hand, look at the bright side. You won't be learning calculus in an auditorium that seats 1000, with a professor who looks to be about an inch high from your seat in the upper balcony.

2.2 What to expect from your instructor

Now that you have an instructor, and you know what he or she is, let's move on to the more refined topic of who he or she is. Here are a few examples of what you might expect.

Famous Mathematician Story   John Von Neumann was a Hungarian mathematician who came to the United States in the 1930s and in his spare time invented the concept of computer programming. He was also a little unusual.

One time a student went up to him after a calculus lecture. "Professor Von Neumann," the student said, "I don't understand how you got the answer to that last problem on the board." Von Neumann looked at the problem for a minute and said, "e^x." The puzzled student thought he had been unclear. "I know that's the answer, Professor Von Neumann. I just don't see how to get there." Von Neumann looked at the student for a minute, stared into space, and repeated, "e^x". The student started to get frustrated. "But how did you get that answer?" Von Neumann turned to the student and said, "Look kid, what do you want? I just did it for you two different ways."

Moral   Sometimes professors have a hard time remembering what life was like before they knew calculus inside out. Having taught the same material over and over again, year after year, they just don't understand why the students haven't mastered it yet.

Famous Mathematician Story   Norbert Wiener was perhaps the greatest U.S. mathematician in the first half of the twentieth century, revered among his colleagues for his brilliance. He was also famous for his absentmindedness.

After a few years at MIT, Wiener moved to a larger house. His wife, knowing his nature, figured that he would forget his new address and be unable to find his way home after work. So she wrote the address of the new home on a piece of paper that she made him put in his shirt pocket. At lunchtime that day, the professor had an inspiring idea. He pulled the paper out of his pocket and used it to scribble down some calculations. Finding a flaw, he threw the paper away in disgust. At the end of the day he realized he had thrown away his address. He now had no idea where he lived.

Putting his mind to work, he came up with a plan. He would go to his old house and await rescue. His wife would surely realize that he was lost and go to his old house to pick him up. Unfortunately, when he arrived at his old house there was no sign of his wife, only a small girl standing in front of the house. "Excuse me, little girl," he said, "but do you happen to know where the people who used to live here moved to?" "It's okay, Daddy," said the little girl. "Mommy sent me to get you."

P.S.: Norbert Wiener's daughter was recently tracked down by a mathematics newsletter. She denies he forgot who she was but admits he didn't know his way to the house.

Moral 1   Don't be surprised if the professor doesn't know your name by the end of the semester.

Moral 2   Be glad your parents aren't mathematicians. If your parents are mathematicians, introduce yourself and get them to help you through the course.

Famous Mathematician Story   David Hilbert was one of the great European mathematicians at the turn of the century. One of his students purchased an early automobile and died in one of the first car accidents. Hilbert was asked to speak at the funeral. "Young Klaus," he said, "was one of my finest students. He had an unusual gift for doing mathematics. He was interested in a great variety of problems, such as . . ." There was a short pause, followed by, "Consider the set of differentiable functions on the unit interval and take their closure in the . . ."

Moral 1   Sit near the door.

Moral 2   Some mathematicians can be a little out of touch with reality. If your professor falls in this category, look at the bright side. You will have lots of funny stories by the end of the semester.

How to deal with your instructor

Here are the basic rules of etiquette for successfully interacting with your instructor.

1. Know your instructor's name. Many math instructors can't imagine that there is anyone else in your life of greater importance. Play to that. Use it to your advantage.

2. Decide what title to use when addressing your instructor. This apparently delicate question of etiquette is easily solved by using gross flattery. Suppose that your instructor's last name is "Gnyrd," for example. Then call him Dr. Gnyrd. Even though his earnings will never approach those of his pre-med students, his ego will be entranced by this distant link to the medical profession. If your instructor is female, call her Dr. Gnyrd. It is inexpressibly annoying to female professors to have students call them Ms. or Mrs. Gnyrd when they hear male graduate students obsequiously addressed as Dr. Gnyrd. Be generous with titles. No matter what they say to the contrary, no instructor in any subject will ever be offended by use of the title "Doctor." While this designation is inappropriate for graduate students who have not yet completed their doctorates, they'll still be thrilled to pieces.

3. Avoid rudeness. Remain polite at all times. You can be firm in seeking out explanations, but do not be rude. Remember, your instructor ultimately determines your grade. Even the most idealistic professor, who would never let his or her personal dislike of a student affect that student's grade, may find that at the end of the semester the student is dead center on the borderline between two grades. Consciously, the professor says, "Well, that student did a little worse on the final than on the previous exam, and since the final is cumulative, that justifies my giving the lower grade," while subconsciously, the professor is saying, "I am going to nail that nasty little kumquat to the wall." Though grading mathematics is not as arbitrary as grading your course Iconoclastic Modernity in the Hermeneutic Tradition, there is still some leeway available. So stay on the professor's good side.

General Principles of Acing Calculus

1. Buy this book. If this is not your copy, get your own. Keep it next to your bed. Maybe buy a couple of extra copies for the kitchen and bathroom.

2. Pick the right professor.

3. Pay attention. Easier said than done. Some calculus professors spend years refining techniques to make your attention wander and your mind dip into numbness. Unless you have a brain of steel you will be unable to withstand these assaults. Discretion is the better part of valor here—choose the right instructor if you have any choice. If not, find out where the really strong coffee is sold. Every college town should have its sources of pure caffeine. It it doesn't, forget calculus and open a cappucino bar. You could make a fortune.

   One good way to force yourself to pay attention is to sit in the front row. Experience has shown that those who intend to ace calculus either

   ✳ Sit in the front row, where attention is maximized, or

   ✳ Sit in the back row and don't pay any attention. These people have already taken the course in high school and are retaking it for an easy A. They will not buy this book. Little do they know that readers of this book are going to dominate the top of the grading curve, knocking them down to a B (if they're lucky).

4. Do the homework problems. Despite the fact that most of the focus of university instruction is in the classroom, most of the learning takes place when you sit in the quiet of your bedroom, or the library, or a jacuzzi, and do your homework problems. If you understand how to do homework problems, then you understand how to do exam problems, too.
    Here is a simple plan for doing homework problems.

    * See which problem you are assigned.

    * Look for an example in the book that looks sort of similar.

    * See if what is done in the example works for your problem. Most of the time it will.

    After you do this a few times you won't need to look at the examples anymore. You now know how to solve this problem. That's all there is to it.

5. Get help. When you are stuck on a problem, do not

    * Bang your head against the wall.

    * Decide that a couple of beers would help.

    * Give up in disgust, and resign yourself to a career in Slurpee sales.

    Instead, get help. Who to see? In descending order:

    * The sharp people in the class. Swallow your pride. Pride won't get you that A. Someone else can explain the problem. And remember, don't ask for the answer, ask for the explanation. The answer alone will do you no good on an exam when the problem has been changed slightly. Get a relationship going with these people, a give-and-take: they give, you take. It's actually good for them, too. There is no better way to learn something than having to explain it to somebody else.

    * Somebody who took the class recently (and did well). If it's been awhile, though, you will spend more time reminding the person what calculus is about than getting help.

    * The teaching assistant. Often, having begun teaching only a semester or two before, teaching assistants are still extremely enthusiastic about spending hours of their time with students. They are power happy and revel in the fact that they know more than you. Take advantage of it.

    * The professor. Unfortunately, the few office hours available are often inconveniently scheduled. So for this you have to plan ahead. What is the best procedure for visiting the professor during office hours? Come

in well prepared. Use those stickum notes to write down questions and stick them in your homework and lecture notes in the right places. Then run down the questions with the professor, one after the other. The professor will be so impressed with your organization and concern for the course material that you could improve by an entire grade at the end of the course. Of course, it helps if the professor knows your name.

Most professors are more than willing to make appointments to see students outside office hours. There are a few exceptions. One student cornered a professor after class and requested an appointment to clear up some confusing points. The professor looked in her appointment book and said, "I have an opening in two and a half months, on March 24 at 6:15 A.M." "What!" the student gasped. "That's the best you can do? That's the only time you can see me?" "That's it," said the professor. "What's wrong, do you have a problem with that?" "Well, you see," said the student, "that's when I'm scheduled to see the teaching assistant."

✳ Email. Email is the easiest and fastest way to get an answer out of an instructor or a teaching assistant. Surprisingly few students use email to communicate with faculty. It is particularly useful for short questions like "Is class going to be canceled again tomorrow?" and "Can I make an appointment to see you Tuesday?" Professors check their email all the time and usually are quite willing to respond to it. You can send a message to them at any time of day or night. They vastly prefer it to that 3:00 A.M. phone call to their home asking for the homework assignment.

✳ The web. The World-Wide Web has lots of calculus sites and calculus discussion forums, where you can get answers to various questions. So if you know how to surf the web, catch that next wave. If not, look for the student in your class who has gotten the least sun and ask him or her to teach you how.

6. Know the examples. Calculus is usually presented as a bunch of rules (sometimes organized as theorems and lemmas) with an occasional example given to illustrate them. Pay attention to those examples! You may have the idea that they're randomly picked from an ocean of possibilities. However, the collection of examples is more like a small pond, so you should be particularly interested in the ones your instructor presents. Chances are that problems appearing on finals won't be very different from examples seen in class. Mathematicians recycle exam problems in much the same way that comedians recycle jokes.

7. Get hold of old exams. You may think there is something unethical about using previous exams as a study aid. In fact, previous exams are almost always in the public domain. At many universities they are even available for examination and copying either in the library or at the offices of the

mathematics department. If they aren't, seek out students who have taken the course previously, visit instructors who taught the course and ask them for copies, do whatever you must. Get them. Do them. Understand them. Another valuable source of sample problems is other calculus texts. Used bookstores have plenty of texts from two or three years ago that sell for pennies on the dollar. The mathematics in these books does not become obsolete in five years (or even 50), and mathematics instructors are notorious for taking exam problems from other texts.

8. Study. Studying will be unattractive to some of you, unfamiliar to others, but the truth is that it cannot be avoided in the end. The purpose of this book is to show you how to study effectively and in a way that is fun, not how to avoid studying altogether. If you cannot come to grips with this, you might consider a career in politics or some similar field that does not require mathematics or higher-level thought.

9. Do bonus problems. If you are lucky enough to have a professor who gives bonus problems, then whatever else you do, *do the bonus problems!* Why? Because almost no one else will do them. Many students have a misconception about bonus problems. They think that in some vague, unspecified way, bonus problems don't count. But they do. They get added to your total for the semester. If no one else does them, your total score for the semester is that much higher than anyone else's, meaning you get the high grade, and everyone else gets to sell magazine subscriptions door-to-door for a living.

10. Avoid the dark side. Almost without exception, cheating does not lead to higher grades. You might be able to squeak through one problem here or there, but pretty soon you will be in the middle of material that depends on material previously covered. Your familiarity with the previous material will be what you glimpsed on your neighbor's test during the last exam, and let's face it, a "33" half covered by a hand is not enough background on limits to understand derivatives. Cheating is high risk with no reward. (Now, we know what you're thinking. Of course a bunch of professors are going to say that. They are required to say that by some sort of Newtonian oath they took when they signed on to be profs, the same oath that says they have to erase the board for the next professor at the end of each class. But in fact, there are no such provisions in the oath. Cheating really is more likely to flush you down the tubes than to help you. And we erase the board only if we like the next professor.)

Good and Bad Questions

4.1 Why ask questions?

What? Me ask questions? No way! That's dangerous. What if I ask a stupid question and look silly, ruining my social life? What if the professor says, "You idiot. How can you ask such a foolish question? Get out of this class and take this F with you to use as a grade"?

The vast majority of students feel this way. Coincidentally, the vast majority of students get grades B, C, D, and F. In fact, asking questions will never get you in trouble, and it will usually score you some points. Professors love questions because questions make it appear that the class is engrossed in the lecture, that the students and the professor are in synch, communicating, interacting; learning is happening; all is good.

Students like questions because they get a chance to recover their lost attention span, hear what's troubling the other students, and, most important, see what the questioner is wearing.

Questions asked in class can serve various aims:

∗ To fill class participation requirements. Sometimes class participation is used in the grading process. In this case it is essential to have a store of appropriate questions to get your participation points. Otherwise you may be put on the spot by having to come up with answers, which can be risky.

✱ To get the instructor to explain something. Instructors will occasionally clear up a confusing point if you ask an appropriate question. Depending on the class size and the temperament of the instructor, this may or may not be a good strategy.

✱ To impress the person sitting next to you. You meet a fine class of people in calculus, and nothing impresses a potential date like a mastery of the techniques of integration.

Remember that timing is everything. If you use questions like the ones in Section 4.2, be sure to choose an appropriate time.

 ## Some sample questions

Here are some questions that might be appropriate models. More will show up later in the topics sections of this book.

1. What does the word "calculus" mean?

   Wrong answer: The layer of cavity-causing deposit that builds up on your teeth when you don't brush.

   Right answer: A system of mathematical results going back to Newton and Leibnitz used for calculating slopes and areas.

2. Where did you get those ultracool shoes?

   Better to butter up the professor early, before someone else beats you to it.

 ## Questions not to ask

1. Is this good for anything?

   Asking a professor this question is like asking the guy who cleans up after the elephants, "How come you wear those silly rubber boots?" To them, the answer is so obvious, the question hardly makes sense. The basic theories of electricity, light, sound, matter, population growth, economics, epidemics, statistics, and stamp collecting, just to name a few, all depend on calculus. Without calculus, economists could not make pinpoint accurate forecasts, and weather prediction would not be the impeccable science that we have grown to expect. TVs would explode, airplanes would fall out of the sky, and athlete's foot would be an incurable disease.

2. Questions that were just asked.

   You will impress nobody with a question about something that was discussed just minutes ago, while you were jabbering with your friends about last night's big party or before you showed up late to class.

3. Will this be on the exam?

   Some students repeatedly ask this question throughout the semester. This is dangerous, since by the time exam time comes around the professor will probably have forgotten what was and what was not supposed to be on the exam. It also convinces the professor that you could care less about the material in the course and that all you care about is your grade. This hurts the professor's feelings, and when people's feelings are hurt, they tend to lash out at the perpetrator. So regardless of whether it's true that all you care about is your grade, don't let on.

4. Is the fig newton named after Isaac Newton?

   Isaac Newton had one of the greatest minds in human history, sharing only with Karl Friedrich Gauss the honor of making the all-time top three list in both the mathematics and physics categories. But clever about cookies he was not. All of his attempted cookie recipes were disasters. Some blame his pecan sandies for causing the black plague, but this is probably unfair. (Actually, the fig newton was invented in Newton, Massachusetts, which was in turn named after old Isaac, so indirectly the answer to the question is yes.)

CHAPTER

5

Are You Ready? Calc Prereqs

It's a good idea to spend a few minutes reviewing what you're supposed to have learned in high school in all those classes meant to prepare you for your first calculus class.

5.1 What you think you learned

* Up to ninth grade: Really complicated arithmetic, including elaborate long-division problems, which everyone knew could be done more easily on a calculator.

* Tenth through twelfth grades: Well, algebra, geometry, and trigonometry, right?

Algebra. How to solve complicated questions involving water going upstream, boats going downstream, and trout fishing, by throwing in a lot of x's and y's and hoping some plausible numbers come out at the end, like they caught four trout apiece and were moving at a rate of 7 miles per hour, or vice versa.

Geometry. How to show that one line that was incredibly far away from another line, barely on the same page even, was actually parallel to the first line. Why anyone cared about this escaped you at the time, but on the other hand, what was the point of dissecting the frogs in biology?

Trigonometry. Hmmm, missed that altogether, did we? But it's worth mentioning separately because it's the first math word, coming well before "calculus," that has the capacity to intimidate other people.

5.2 What you really need to know on the first day of class

ALGEBRA

1. Be able to factor something like $x^2 - 6x + 8$ into $(x - 2)(x - 4)$.

2. Be able to find the values of x that satisfy $x^2 - 7x + 9 = 0$ using the quadratic formula.

3. Know that $x^2 - y^2 = (x - y)(x + y)$ and that $x^2 + y^2$ doesn't factor.

4. Know that $\sqrt{x^2 + 4}$ DOES NOT EQUAL $x + 4$ OR $x + 2$.

5. Know that $(9x)^{1/2} = 3\sqrt{x}$.

6. Know that $\dfrac{x^2 x^8}{x^3} = x^{2+8-3} = x^7$.

7. Be able to find which values of x satisfy $\dfrac{x - 2}{x + 4} < 7$.

If you are mystified by some of these topics, you will need to fill in the gaps in your knowledge before you can expect to be a master of the calculus universe.

FUNCTION NOTATION

Throughout calculus, functions are denoted by $f(x)$ as in

$$f(x) = x^2 - 7x + 5$$

What the "(x)" indicates is that x is the quantity that varies. We fondly call x the "variable." It plays the role that the coin slot plays on a vending machine at a highway rest stop. You put 75 cents in the vending machine, and out comes a comb. You put in $1.00 and out comes a pack of playing cards. Same with the function f, except no comb or playing cards. You put in 2 for x and you get $f(2) = (2)^2 - 7(2) + 5 = -5$. You put in 3 for x and you get $f(3) = (3)^2 - 7(3) + 5 = -7$.

ABSOLUTE VALUE FUNCTION

This is one of the truly misunderstood functions of our time. It doesn't deserve its bad reputation.

$$f(x) = |x|$$

What could be more innocuous? It's just an x with two little bars on either side. The most important thing is to know the actual definition of the absolute value function. Many people just remember it intuitively, saying to themselves, "Well, it just means make x positive." That's true, but unless it's applied carefully, it can lead to trouble.

Officially,

$$|x| = \begin{cases} x & \text{if } x \geq 0 \\ -x & \text{if } x < 0 \end{cases}$$

That's it. So $f(3) = |3| = 3$ and $f(-2) = |-2| = 2$. No problem.

But what do you do if $f(x) = |x - 2|$? This may make you nervous. But there's no need to break out in hives, because we know the definition of the absolute value function, so we know exactly what to do. Let's just write out what the definition says.

$$|x - 2| = \begin{cases} x - 2 & \text{if } x - 2 \geq 0 \\ -(x - 2) & \text{if } x - 2 < 0 \end{cases}$$

Let's write it one more time, slightly simplified.

$$|x - 2| = \begin{cases} x - 2 & \text{if } x \geq 2 \\ -(x - 2) & \text{if } x < 2 \end{cases}$$

Not so bad. Suppose that we wanted to graph it. Then we would graph $y = x - 2$ for $x \geq 2$ and we would graph $y = -(x - 2)$ for $x < 2$. And voilà, Figure 5.1.

If you want to show off, you can define $|x| = \sqrt{x^2}$. This definition is equivalent to the one we gave above because x^2 is always positive and the square root of a positive number always refers to the positive square root. But this definition can be confusing if you don't have an irresistible affinity for square roots.

GEOMETRY

Geometry played an important role in teaching you about the forms of a mathematical proof, assuming you weren't too busy passing through puberty

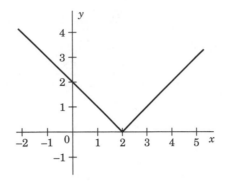

Figure 5.1 Function $f(x) = |x - 2|$.

to pay attention. We will quickly recap what you really need to know in Chapter 7, "Lines, Circles, and Their Friends." As to all the stuff about intersecting bisectors and the side-angle-side theorem, you can leave it to Euclid. We won't need much of it. If you come away with some idea of what a proof is and what similar triangles are, you're ahead of the game.

TRIGONOMETRY

Figure 5.2 contains most of what you need to know.

$$\sin \theta = \frac{\text{opposite}}{\text{hypotenuse}}$$

$$\cos \theta = \frac{\text{adjacent}}{\text{hypotenuse}}$$

$$\tan \theta = \frac{\text{opposite}}{\text{adjacent}} = \frac{\sin \theta}{\cos \theta}$$

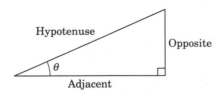

Figure 5.2 Three sides of a right triangle with one angle equal to θ.

$$\csc \theta = \frac{\text{hypotenuse}}{\text{opposite}} = \frac{1}{\sin \theta}$$

$$\sec \theta = \frac{\text{hypotenuse}}{\text{adjacent}} = \frac{1}{\cos \theta}$$

$$\cot \theta = \frac{\text{adjacent}}{\text{opposite}} = \frac{1}{\tan \theta}$$

We can measure an angle in either radians or degrees. One full circle can be described as having 360 degrees or 2π radians. Since $360° = 2\pi$ radians, dividing both sides by 360 shows that $1° = \pi/180$ radians.

You should know the values of all of the trig functions at the standard angles of $0°, 30°, 45°, 60°, 90°$, and $180°$. You should also know what these angles are in radians.

$$1° = \pi/180 \text{ radians}$$

$$30° = \pi/6 \text{ radians}$$

$$45° = \pi/4 \text{ radians}$$

$$60° = \pi/3 \text{ radians}$$

$$90° = \pi/2 \text{ radians}$$

$$180° = \pi \text{ radians}$$

$$360° = 2\pi \text{ radians}$$

Converting from degrees to radians is easy. You multiply by $\pi/180$ to go from degrees to radians, and by $180/\pi$ to go from radians to degrees.

Two triangles are helpful here, the 30°-60°-90° triangle and the 45°-45°-90° triangle, shown in Figure 5.3. Facts like $\sin(30°) = 1/2$ can be read right off these triangles.

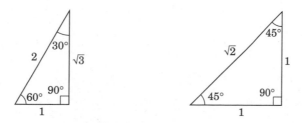

Figure 5.3 Two triangles to remember.

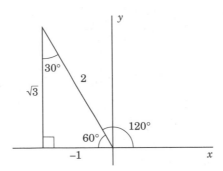

Figure 5.4 Finding trig functions of 120°, or $2\pi/3$ radians.

If you need to find the trig functions at some angle larger than 90°, like 120°, that's no problem at all. First draw the x and y axes and use the fact that 120° is 60° less than 180°. So draw a 30-60-90 triangle on top of the negative x-axis, with the 60° angle at the origin. See Figure 5.4.

Labeling the bottom edge with a minus sign, since it is along the negative x-axis, we have $\sin(120°) = \sqrt{3}/2$ and $\cos(120°) = -1/2$.

There are a variety of trigonometric identities, but almost all the information you're likely to need is in one, namely,

$$\boxed{\sin^2 x + \cos^2 x = 1}$$

That one is worth remembering.

We will list a few others, just for fun and in case you want to refer to them later:

$$\sin(2a) = 2\sin a \, \cos a$$

$$\cos(2a) = \cos^2 a - \sin^2 a = 1 - 2\sin^2 a$$

$$\sin(a + b) = \sin a \, \cos b + \cos a \, \sin b$$

$$\cos(a + b) = \cos a \, \cos b - \sin a \, \sin b$$

$$\cos^2 a = \frac{1 + \cos(2a)}{2}$$

$$\sin^2 a = \frac{1 - \cos(2a)}{2}$$

You may eventually run into the infamous "inverse trig functions" arcsin, arccos, arctan, and the other members of the arc family. The meaning is simple: If $y = \sin x$, then we say $x = \arcsin y$. For a particular example, since $\sqrt{3}/2 = \sin(60°)$, we can calculate that $\arcsin(\sqrt{3}/2) = 60° = \pi/3$ radians. And that's all we need to know about inverse trig functions for now.

COMPOSING FUNCTIONS

Beethoven composed symphonies and went down in history. In this section we'll show you how to compose functions. Probably history won't blink, though it may give a chuckle or two. Just goes to show, "You can use the same name, but that don't make it the same."

Composing two functions means taking the variable x, applying one function to it, and then applying another function to the result. So if we take the functions

$$f(x) = \sqrt{x}$$
$$g(x) = x + 7$$

we can *compose* them by, say, first doing f and then g. We'd write

$$g(f(x)) = g(\sqrt{x}) = \sqrt{x} + 7$$

We just replaced each x that occurred in the function $g(x)$ by a $\sqrt{x}$.

Or we could first do g and then f:

$$f(g(x)) = f(x + 7) = \sqrt{x + 7}$$

Here, we replaced each x that occurred in the function $f(x)$ by an $x + 7$.

Notice that we get two different functions, so *the order is important.* The order that we use to calculate the new function is the opposite of what you ought to have been taught about using multiple forks in elegant restaurants: While you're supposed to start with the outermost fork and proceed inward till dessert, here you start with the innermost function and proceed outward. For those of you who have never learned how to eat in an elegant restaurant, learn calculus and make oodles of money. Then you can go to the most expensive restaurants and use the forks in any order you want.

Here's a trick if you get confused. Let's take two new functions

$$f(x) = \cos x$$
$$g(x) = x^2 + x$$

and calculate the function $g(f(x))$.

First take the function g, and everywhere you see an x on the right-hand side, put parentheses around it:

$$g(x) = (x)^2 + (x)$$

Think of the x's in parentheses on the right-hand side as being *exact copies* of the (x) you see on the left-hand side. This means that whatever gobbledygook you see in the parentheses on the left-hand side has to be reproduced exactly wherever you see (x) on the right-hand side. So,

$$g(\text{gobbledygook}) = (\text{gobbledygook})^2 + (\text{gobbledygook})$$

In our case,

$$g(f(x)) = (f(x))^2 + (f(x))$$

Finally, take equal signs very very seriously; you know what $f(x)$ equals, so everywhere you see $f(x)$ replace it with $\cos x$:

$$g(f(x)) = g(\cos x) = (\cos x)^2 + (\cos x)$$

Kind of scary that all the math you learned in the last five years of your life can be distilled to five pages. But then, we did leave out lots of important material that just happens to be not relevant to calculus. Take the side-angle-side theorem, for instance. Now aren't you glad you learned that? And of course there was that unit in eighth grade on how to figure out the cost of a loose-leaf binder on sale for 30% off. So don't worry. Those years were chock full of useful tidbits.

Besides, this was the period in your life when you had to deal with the onset of acne, hormonal tidal waves, and the overriding question, confronted every day: Would anyone sit with you at lunch? It's amazing your teachers got anything to stick.

5.3 Computers and calculators: Our 2-bit friends

These days, chances are good that your calculus course will be using computers (running a computer algebra system), a graphing calculator, a programmable calculator, or perhaps regular old calculators. If so, you have entered the world of digital calculus. These tools can make your life easier when you do calculations and can increase the power and scope of your knowledge. They can allow you to experiment quickly with many examples, giving you a better understanding of the subject. On the other hand, they can lead to frustrating moments when you have trouble finding the on/off switch or the alt-3 keys. So let's talk about the various possibilities in turn.

First, computer algebra systems. There are several common systems that are used in calculus. They go by the names Mathematica,
Mathcad, Maple, and DERIVE. Each is a computer algebra system, a
software package that juggles formulas and equations. These programs can
do algebra, like factor $x^2 - 5x + 6$ or simplify the fraction $\dfrac{x^2 - 1}{x - 1}$. That's a
nice feature to have. But in fact, they can do a lot more than that. They
can do many of the basic calculations of calculus: take derivatives of simple
functions (and not-so-simple functions), and determine indefinite and definite
integrals. These are operations that we will be talking about throughout this
book, and they form the basis for calculus. Sophisticated calculators can also
do these basic operations.

You may be tempted to ask, "Wait a minute. If these machines can do
calculus, why do I have to learn it? Let them do it, and I'll just hang out at
the pool and drink brightly colored tropical drinks."

But of course, the point is that the computers don't understand what they
are doing. They cannot interpret their results and use them to make bundles
of money the way you can. And a good thing too, or they'd be hanging out by
the pool and you'd be spending all your time learning how to polish memory
chips.

One place that computers and calculators do a great job is in graphing
functions. After you put in a function, they can graph it, display it from all
sides, and determine where its peaks and bottoms occur. Combined with a
solid understanding of techniques of graphing, you will become a graphing
grandmaster. You can graph GNP, GPA, and MTV.

The worst thing about computers and calculators is figuring out how to
get started on the little gadgets. Once you get the hang of which keys to
push, they can be tremendously useful. But nothing is more frustrating than
wasting time on the mechanics of the machine instead of applying its power.
Too many students spend an hour trying to find a command, only to find out
they're looking in the wrong manual. So the key is to harness the power of
the machine.

The best way to do this is to get someone else to show you. Latch onto
a fellow student or an instructor or a computer consultant until you can
work through a few example problems. Make sure they show you the whole
thing, from the time you turn on the power until you put the thing away.
With computing, it doesn't help to know 99% of the process for solving your
problem if you don't know which directory your account is in or how to access
a program. So have someone else go through it all with you first.

As you progress with computation, the time will come when the machine
decides to test you. It will lose your files or ignore your keystrokes or otherwise
test your sanity. The usual response to this is "Oh, no, I did something wrong,
I'm going to get in trouble and fail the class. Help!" The correct response is "I
bet that cute cyber-star over there would love to come and help me figure out

what hung my computer." Consider these situations a chance to experiment with your social skills and the computer lab as your own giant petrie dish of opportunities.

There are various other places where computers and calculators are useful in calculus. For instance, they can give a good intuition for limits in general and the limiting process that occurs in derivatives in particular. In Section 22.5 on numerical integration, there are large quantities of numbers to add together. We have included short programs at the relevant points in the book that can be utilized on a programmable calculator or computer to do calculations.

Calculators and computers can be powerful tools in learning calculus; they offer you possibilities of in-depth understanding that were previously available only to demigods. Using them to explore the manifold wonders of calculus can lead to a deeper grasp of the subject. And remember, knowledge is power and power costs 10 cents per kilowatt.

How to Handle the Exam

Have you ever kicked yourself after an exam and said, "I knew how to do all that." But somehow when confronted with the problems on the exam, you took square roots when you should have squared and used degrees when you should have used radians. What can you do to avoid having this happen again?

Learning math is a lot more like learning a language than most students realize. When you learn a language, you often feel good after you have learned a few key phrases like "Où est la toilette?" Then off you go to France, confident that you are going to walk on French water. But after a couple of weeks of wandering around Paris, where everyone else is eating delicious pastries and gawking at very old paintings and all you can ask for is the nearest latrine, you realize that perhaps a bit more practice in the language would have been worthwhile.

Same with math. Understanding the broad outline isn't good enough. You have to be fluent, conversant to such a depth that you just automatically know what to do. It takes no thought process on your part. It becomes innate.

Of course, in becoming fluent, you want to study the important things first, namely, what will be on the exam.

6.1 What will be on the exam

The main source of information about what will be covered on an exam is the instructor. Probably there is a syllabus stating exactly what will be covered. Possibly the exam topics were announced in class.

But the key period for culling information is the week prior to the exam. This is when the instructor is making up the exam. Before that, the instructor probably has no idea what will be on it. (There are exceptions to this. In particular, there are the instructors who have two exams and use them alternately every other year.)

In fact, instructors hate to give exams. It's a lot of work making them up and pure agony grading them. If they could get students to do the studying without using exams, you'd never have to worry about another exam.

CALCULUS IN THE AFTERLIFE

Two older instructors are sitting around the department lounge, talking about how much they love to teach but hate giving exams. "In heaven, I bet they have no exams!" says one of them. "All the students would come to class prepared, and there would be no exams to give and grade." "What a place," says the second. "Hey, let's make a pact. Whoever dies first comes back and tells the other what it's like to teach calculus in heaven." "Sounds good," says the first. A week later the first instructor slips on a discarded pocket protector and leaves this world. That night, while the second instructor is sleeping, the first one comes to visit him in a dream. "So, what's it like?" asks the second. "Well, I have good news and bad news," says the first. "The good news is that there is calculus teaching in heaven, and I have to tell you, the students are fantastic. Enthusiastic. Attentive. It's a dream come true. And no exams whatsoever! They don't call it heaven for nothing." "Wow, that sounds great," says the second. "What's the bad news?" The first shrugs. "You're teaching on Monday."

As the time for the exam approaches, it is likely that your instructor will realize that some key topic that should appear on the exam was never discussed in class. So drink coffee, stick a tack in your shoe, do whatever it takes to pay close attention during any review sessions held immediately before an exam. Don't let any words go unheeded as the test date approaches.

6.2 How to study

1. Go over the assigned homework problems. Make a list of all the homework problems you were given. Pick one problem at random from each section and give it a whack. If it whacks you back, you need to review that section.

Do this with all the material and the exam is in the bag. It is almost always true that if you have mastered the assigned homework problems you will do well on the exam.

2. Dig up old exams and do them. An opportunity for the enterprising student! It is far easier for an instructor to reuse an old exam than to cook up a new one. Exam questions are really hard for a professor to make up. Questions must be not too easy but not impossible (although this restriction is sometimes inadvertently overlooked). All the numbers have to work out nicely. So the instructor will very likely leaf through the previous couple of years' exams and modify some problems slightly. Not more than two years because the older exams are hopelessly lost under piles of papers.

3. Seek professional counseling. What if you don't know how to do a problem, and you can't make heads or tails out of the convoluted explanation in your textbook, and it's not one of the subjects covered in this trusty book? What then? It's time for the pros to earn their keep. Seek out your professor or teaching assistant for help. Unfortunately at such times they tend to be

   * In bed sleeping, since it's 3:00 A.M.

   * In their offices surrounded by 247 other students who have finally realized that the professor's promise that "eventually, this will all become clear" was a substantial misrepresentation of the truth.

   * Hiding out somewhere. Nowhere to be found.

   So what to do? Panic perhaps? No, no, no. You have not yet used the really valuable resource, your fellow students.

4. Seek amateur counseling. Students who have mastered a topic like integration by parts achieve no greater joy than when explaining it to their unenlightened peers. You can vastly increase the collective happiness and self-esteem on your campus by finding the student who was paying attention the day you slept late and getting her to explain the idea to you. A student who took the course last year might also do the trick, though chances are good that much of what he remembers is wrong. Anyway, student explanations are much less likely to involve phrases like "by completeness and compactness of the unit interval" or "using the third Peano axiom."

5. Practice under exam conditions. It's easy to think you understand the material better than you really do. You can fool yourself into thinking you can solve a problem when you are looking at the answer book or at a worked-out solution. When the exam comes around, your mind is a blank without the example to follow. Test your knowledge of a topic by trying problems under exam conditions. Pick out one problem from each of the

types that are to be covered on the exam. Give yourself one hour to do your ersatz exam and see how it goes. If you can do it under those conditions, the exam should be a cinch.

 How not to study for the exam

1. **Park the markers.** Don't bother reading through your calculus text and highlighting stuff with a yellow marker. A tremendous waste of time. This is not an efficient way to study. It *is* a good way to drift into an afternoon nap. Doing problems is the way to go.

2. **Don't spend a lot of time** bothering your professor with questions about what will be on the exam unless the professor has completely ignored this issue. Professors hate this kind of question, and it almost always fails to provide useful information. What will be on the exam is completely obvious—exactly what was on the exam last year and on the exam the year before. Time spent wheedling is better used studying.

 Taking the exam

1. **Arrive on time.** Excuses for being late are rarely successful.

   A tired excuse: The night before one exam, two students tied one on (well, actually, tied two on, one each) and managed to sleep through the final. They realized they were in serious trouble, so they agreed to tell the professor that they had a flat tire on the way to the exam. "No problem," said the professor, "Come by my office at 5:00 P.M. and I'll give you the exam then." Feeling pretty clever, the students spent the intervening time getting information on the exam from students who had already taken it and making sure they knew how to do the problems. At the professor's office that evening, they were told, "Leave your books in my office, and I'll put you in two separate rooms for the exam." They were both ecstatic to see that the professor had given them the exact same exam taken by the class that morning. However, there was an additional page tacked on at the end, upon which was written, "For 50 percent of the grade, which tire was flat?"

2. **Read the problems.** Make sure you are solving the question that is asked. Many a clever student has given a beautiful and elegant answer to the wrong question because of a careless misreading of a problem.

3. **Do the easy problems first.** A far too common mistake is to tackle the problems in the order in which they are given. This leads students to waste disproportionate amounts of time on the first problem they get

stuck on. Sometimes this occurs on the first page of the exam, a recipe for disaster.

4. Get partial credit. Never, never, never (that's right, never) leave a problem blank. Professors want to give you points. It is a compulsive human instinct to want to give at least a few points. You leave the problem blank, and they can't give you even a single lousy point. The professor's hands are tied, literally tied. Of course, if you are putting things down, it is definitely preferable to put something down that is relevant to the problem. If it is a word problem, draw a picture that shows you understand what is going on. Label the picture with some variables. If it's a graphing problem, draw some axes, label them, and plot a few points. Whatever, just don't ever (that's right, ever) leave a problem blank.

5. Don't erase in panic. Never panic at the last moment and erase a whole page of work in the last five seconds. If you have to panic, just cross out the wrong stuff with a single line. Tender-hearted professors have been known to ignore the line if the work is correct, giving at least partial credit.

6. Don't leave early; check your work. Never, never, never (that's right, never) leave an exam early. Use any extra time to go over the exam. On scrap paper, do each problem over from start to finish. Then compare your new answer with your old answer. If there is a discrepancy, figure out which one is right, and the sooner the better. Also check if your answer seems reasonable. If the problem asked for the number of dogs neutered in Walla Walla in 1993, the answer shouldn't be -4596. As dogs are only too aware, you can't un-neuter a dog.

The only caveat to the "Don't leave early rule" is if your professor was emotionally impacted by the 1960s and gives exams with no time limit. Then when it starts to get dark, and it's only you and the professor left in the exam room, go ahead, pack up your scrap paper and turn in the exam.

7. Go over the exam when it's returned. In a class of 100 students there is a high probability that at least one person has been robbed on some problem. After grading forty-five $\frac{dy}{dx}$'s, a $y'(x)$ can look very strange and be marked wrong. So if you don't understand what you did incorrectly on a problem, by all means ask the professor or teaching assistant. Even if it has been graded correctly, you want to understand your mistakes so that you don't make them again. Don't bother with those classics "I feel I deserve extra points for this problem because I knew how to do it but ran out of time" and "I was thinking the right answer but wrote it down wrong." Only a very inexperienced teacher would fall for this stuff. Others may even scrutinize your exam for other points to deduct. Professors dislike it when students ask for unjustified extra points on problems. Do so at your own risk.

8. Don't beg. Your professor is unlikely to improve your grade because you say your parents will hold up payments on your Beamer if you fail the class. Now it's true that begging might work to a limited extent with some inexperienced or exceptionally soft instructors. But even if it did work, you would always know deep inside that those points weren't really deserved, and eventually the guilt would send you crawling through the door of every sleazy bar you could find until, one day, you would wake up, dirty, disheveled, and distraught, with a pounding head and a tongue of sandpaper, and you would realize you had to go over to the math building and find that professor and say, "Please, take those undeserved points away. I can't stand to have them anymore." And the professor would say, "I don't know you. No one as completely decrepit as you has ever been a student of mine." And the professor would call security and they would throw you back into the nearest gutter off campus. So save yourself the embarrassment.

Lines, Circles, and Their Friends

7.1 Cartesian plane

The *cartesian plane* (aka the *coordinate plane* or the *xy plane*) is the name used to describe the piece of paper you draw your graphs on. It works by labeling points with two numbers, called *coordinates*. This pair of numbers is written $(2, 7)$ or $(3, 12)$, or in general (x, y) with the first one called the x coordinate, the second one the y coordinate. A point is chosen (lucky point) and called the origin. The number x then gives the distance to the right of the origin and the number y gives the distance above the origin. See Figure 7.1.

René Descartes was a mathematician who also dabbled in philosophy ("I think, therefore I am"). He introduced his ideas in the 1600s. (Okay, okay, a quick joke here: René takes a break from his work and goes to a party, something he doesn't usually do. The host says, "René, would you like a drink?" "I think not," he says, and poof, he disappears.)

Perhaps because René was right-handed, the x-coordinate measures the distance to the right, rather than the left, of the origin. Similarly, because he lived in France, rather than Australia, the y-axis points upward rather than downward. The reason that the coordinate system is named cartesian rather than descartesian is that Descartes went by his Latin name Cartesius when it came to academic matters. You too can make up your own Latin name for

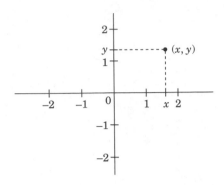

Figure 7.1 Cartesian plane.

use around campus. Just drop a couple of letters at the front of your name, if it seems appropriate, and tack on an "ius" at the end if you're male, an "ia" if you're female. To really impress people, tack on "Caesar" after that. So we're Colinius, Abigailia, and Joelius Caesar.

Until Descartes' discovery, the fields of geometry (triangles, circles, and so on) and algebra (solving equations) were not closely connected. With cartesian coordinates on the plane, an equation like $y = 3x$ becomes a graph, something we can draw and visualize. This was a big breakthrough and generated a lot of hoopla. Of course, a big celebration in the 1600s just meant you wouldn't have to eat turnips that night—you got to feast on hog gristle instead.

7.2 General graphing tricks: The parable of the parabola

Okay, so everybody knows how to graph a function, right? Just pick five points, plug them in for the x values, find the y values, plot the five points, and play connect the dots. If you really want top accuracy, pick 15 points instead.

It's a start. But there are clearly some drawbacks to this method. For instance, in graph (a) of Figure 7.2, which is obtained by connecting some points, we entirely miss the fact that the graph of $1/x$ gets close to $+\infty$ as x approaches 0 on the right and gets close to $-\infty$ as x approaches 0 on the left. We are assuming that nothing interesting happens between the points that we picked. That could be *very* wrong.

Also, suppose what we really want to know is where the peak of a function occurs. We could fool around with plotting points all day, trying to zero in on the peak. With calculus, we will be able to find the exact peak in no time.

There are some general rules that will help you understand what a graph of an unfamiliar function looks like. Let's see how these rules apply to a parabola $y = x^2$ (Figure 7.3).

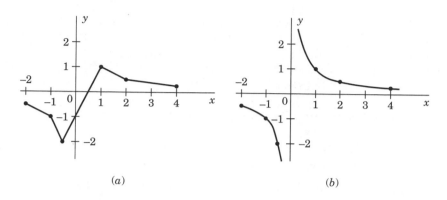

Figure 7.2 (*a*) *Incorrect* result of graphing $y = 1/x$ by plotting some points and connecting them. (*b*) Correct graph.

Parable of the Parabola   Once there was a little parabola curve who, overwhelmed by the ideal images projected by Hollywood and the advertising industry, felt uncomfortable with his body. Bombarded by depictions of beautiful sinuous curves in various states of undress, he felt his shape to be inadequate. In short, he lacked self-esteem. His parents, Mr. and Ms. SugarBowl, sent him in for counseling.

"The problem is, you're ugly," said his therapist. The little parabola was taken aback. "I want a second opinion," he squeaked.

"Very well," said the therapist."Here's a second opinion. You're stupid, too." After pausing to let this sink in, he continued. "Otherwise you would stop worrying about such superficial things as your appearance and concentrate on your inner mathematical soul. Well, we can't do anything about that. But I suppose we can do something about your atrocious shape."

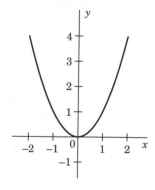

Figure 7.3 Graph of a parabola, $y = x^2$.

So the little parabola got a complete makeover. Multiplied by 5, he became taller and thinner. By adding and subtracting numbers he moved to new neighborhoods and made new friends. After much personal growth, he finally realized his inner child. Unfortunately, the inner child also turned out to be ugly and stupid, but that's another story.

Rules for Moving Around the Graph $y = f(x)$

* Adding a constant to the right side moves the graph up or down. So $y = x^2 + 2$ moves the parabola up by 2 (Figure 7.4a).

* Adding a constant to the variable x moves the graph left or right. So $y = (x - 3)^2$ moves the parabola *right* a distance 3 (Figure 7.4b). Replacing x by $x + 5$ would move the parabola *left* a distance 5.

* Multiply the function by a positive number and the graph loses or gains weight, like a manic depressive on a yo-yo diet. So multiplying by 5 gives $y = 5x^2$, which stretches the parabola upward, turning it from a widespread bowl into a svelte vase (Figure 7.4c). Looking for that new body? Multiply yourself by 5. Go ahead, indulge, multiply yourself by 10—you can never be too thin.

* Multiplying the right side by a negative constant flips the curve over the x axis. So $y = -x^2$ turns the bowl upside down. Now you can use it as a party hat at the calculus party, assuming all the lampshades are taken.

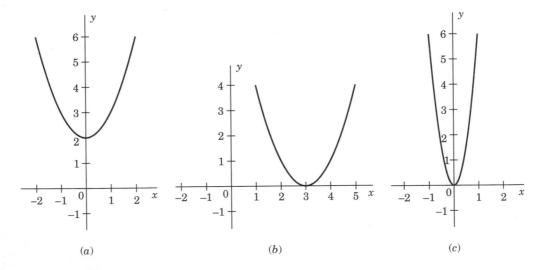

(a)   (b)   (c)

Figure 7.4 Parabolas (a) $y = x^2 + 2$, (b) $y = (x - 3)^2$, (c) $y = 5x^2$.

7.3 Lines

Lines play a very important role in our lives. Where would we be without the classics: "Come here often?" "Haven't we met somewhere before?" "Wow, are you buff. You must work out three hours a day." Then there's the old favorite that Humphrey Bolic tried on Integrand Bergman in the famous math movie *Calcablanca,* "Here's looking at Euclid."

To create many more lines, one needs to use a formula, such as "Why don't you come over to x and I'll show you my y, where x = one of {my room, my tree house, Poughkeepsie} and y = one of {macrame collection, blindfolded warthog imitation, cannelloni}. The x and y are variables that can take on a variety of values. We call this a formula for a line.

It's much the same thing in math. The formula for a line has an x and a y in it, and no scary words (no tan, no sin, no paradigm). The x and y look perfectly normal—they're not living in some seedy denominator somewhere, and they are not raised to any power.

$$x = 3 \qquad \text{GOOD}$$

$$x + y = 3 \qquad \text{GOOD}$$

$$x = 1 - y \qquad \text{GOOD}$$

These give equations for three different lines.

$$e^y = \sqrt{x}, \qquad \text{BAD}$$

$$y = e^x + \arctan x \qquad \text{BAD}$$

These are not equations for lines.

$$x + 4 = y - 3 \qquad \text{GOOD}$$

Yes! A line.

$$xy = x + 1 \qquad \text{BAD}$$

Not a line. Although not too bad, the xy term is sufficiently deviant to stop this from being a line.

$$\frac{2 + x}{y} = 3 \qquad \text{GOOD}$$

A line in hiding! The terms can be rearranged so the equation becomes a more comforting

$$2 + x = 3y$$

Let's take an example, and answer every possible question about it.

Example $4x - 2y = 2$

First Question Is $(1, 2)$ on the line?

Sorry, but no way, since $4(1) - 2(2) \neq 2$.

Next Question How do we find a point on the line?

Well, let's do this the easy way. Let's see where the line intersects the x-axis and y-axis. These points are called the x-intercept and y-intercept, respectively. The beauty of the x-intercept and y-intercept is that for each case, one of the two coordinates is zero. The y-intercept is the point $(0, y)$, where y is the number we need to make the coordinates of the point satisfy our equation:

$$4(0) - 2(y) = 2$$

$$2(y) = -2$$

$$y = -1$$

So the y intercept of our line is $(0, -1)$.
To find the x-intercept, just let $y = 0$ and solve for x:

$$4(x) - 2(0) = 2$$

$$4(x) = 2$$

$$x = \tfrac{2}{4} = \tfrac{1}{2}$$

So the x-intercept of our line is $(\tfrac{1}{2}, 0)$.
We can now draw a flawless picture of our line, since we know two points on it (Figure 7.5).

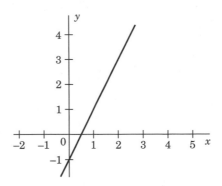

Figure 7.5 Line $4x - 2y = 2$.

Important Note   When graphing a line, it doesn't make the slightest difference which two points you pick to graph it (as long as they satisfy the equation). So if instead, by magic or by some other calculation, we notice that the points $(1, 1)$ and $(3, 5)$ satisfy the equation and graph the line connecting those two points, WE GET THE SAME LINE. Stunning, isn't it?

SLOPES OF LINES

Next Question   What's the slope of the line?

By the slope of a line, we mean a measurement of the tilt of the line. To see how tilted a line is, we pick a random horizontal distance, called the *run,* and find the change in height of the line over that horizontal distance, called the *rise*. Then the ratio, the rise over the run, is a measure of the slope of the line. This terminology was developed by the ancient Egyptians, who had to do a lot of running each year when the Nile would flood and the water would rise. They would mentally calculate the rise over the run so they could get in good enough shape in the preceding months to prevent drowning. Finally, someone had the brilliant idea of constructing the pyramids, so everyone could climb up there and wait for the water to recede without having to hold their breath for three months.

If (x_1, y_1) and (x_2, y_2) are *ANY* two points on the line, then the slope, equal to the rise over the run, will be

$$\text{Slope} = \frac{\text{rise}}{\text{run}} = \frac{y_2 - y_1}{x_2 - x_1}$$

For our example, we'll pick

$$(x_1, y_1) = (\tfrac{1}{2}, 0)$$

and

$$(x_2, y_2) = (0, -1)$$

Then the slope is

$$\frac{y_2 - y_1}{x_2 - x_1} = \frac{(-1) - 0}{0 - (\tfrac{1}{2})} = \frac{-1}{-\tfrac{1}{2}} = \frac{1}{\tfrac{1}{2}} = 2$$

Let's try it with two other points on the line, $(1, 1)$ and $(3, 5)$:

$$(x_1, y_1) = (1, 1), \ (x_2, y_2) = (3, 5)$$

The slope is still

$$\frac{5 - 1}{3 - 1} = \frac{4}{2} = 2$$

Thank goodness.

The slope is traditionally denoted by m, from, "My oh my, the list of synonyms for slope includes hill, incline, obliquity, cant, diagonal, slant, and tilt, none of which begin or even contain the letter m."

Occasionally, the answer to the question "What's the slope of the line?" is "I dunno" or, more formally, "The slope does not exist. There is NO slope to this baby." To see why, take a piece of paper and draw an x-axis and a y-axis. Now draw any other straight line on your paper. For most of you, the line you draw will intersect the x-axis exactly once (the intersection may only be spiritual, that is, off the page) and the y-axis exactly once. Those of you who got up on the wrong side of the bed, however, will draw a line parallel to one of the axes. Those having a really bad day will draw one of the axes themselves. We will call these lines "intercept-challenged" and give them some special attention. The lines parallel to one of the axes will have only one intercept (with the other axis), and funny-looking equations, like $3y = -14$.

In particular, one of the variables takes a brief vacation. Whichever variable is on holiday is the one to which your line is parallel. So the line $3y = -14$ is parallel to the x axis. You can "see" this from the equation once you realize that this equation doesn't give a damn what the x coordinate of a point is; all it cares about is that the y coordinate satisfies $3y = -14$, or $y = -\tfrac{14}{3}$.

We're not even going to discuss the axes themselves, they're too sensitive.

EQUATIONS OF LINES: STANDARD FORMS

There are two basic forms for the equations of lines:

1. Point-slope form:

$$y - y_0 = m(x - x_0)$$

The line has slope m and passes through the point (x_0, y_0).

2. Slope-intercept form:

$$y = mx + b$$

The line has slope m and y-intercept b. The letter b comes from the expression, "Boy oh boy, look where this baby crosses the y axis."

So a line with equation $y - 2 = 4(x - 3)$ has slope 4 and passes through the point $(2, 3)$. Cleaning the equation up by algebra, it becomes $y = 4x - 10$, meaning the slope is still 4 (it had *better* still be 4), and the y-intercept is -10.

7.4 Circles

Circles play an important role in many circumstances. Take for instance the *circle of life*. Now, we're not exactly sure what that is, but it sure sounds important. Or take your *circle of friends*. What would you be without friends? Well, friendless, of course. But back to the topic at hand. Let's say we want to find the equation of a circle of radius r centered at the point (a, b) in the xy-plane. Then every point (x, y) on the circle has a fixed distance of r from the point (a, b). But the distance from (a, b) to (x, y) is $\sqrt{(x - a)^2 + (y - b)^2}$. So the point (x, y) must satisfy

$$\sqrt{(x - a)^2 + (y - b)^2} = r$$

So

$$(x - a)^2 + (y - b)^2 = r^2$$

is the equation of a circle of radius r centered at (a, b), as in Figure 7.6.

And that is all there is to circles. Well, almost all. What if a big grizzled guy with a desperate look in his eye and a doberman pinscher on a leash hands you a piece of paper upon which is written

$$x^2 - 4x + y^2 = 6$$

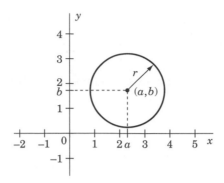

Figure 7.6 The circle of radius r centered at (a, b).

and says, "Tell me what this is the equation of, or Ralph here treats you like a fire hydrant"?

It sounds unlikely, but this happens in Central Park all the time. The equation looks kind of like a circle, but something is wrong. Why not split the equation up into the terms involving x and the terms that do not involve x?

$$(x^2 - 4x) + y^2 = 6$$

Notice that if we add a 4 inside the parentheses, it becomes a perfect square. Of course you can't just add a 4 to one side of an equation. Ralph has relieved himself for less. You also have to add a 4 to the other side of the equation:

$$(x^2 - 4x + 4) + y^2 = 6 + 4$$

$$(x - 2)^2 + y^2 = 10$$

This is a circle of radius $\sqrt{10}$ centered at the point (2, 0).

And that is the sum total of what you need to know about circles, at least as far as Ralph and calculus are concerned.

7.5 Ellipses, parabolas, and hyperbolas

They call themselves the Conical Sections Gang, because they all come from slicing a cone with planes at various angles. Although sometimes intimidating when they approach as a group, they are completely harmless when met one on one, and they often revert to blubbering apologies if confronted over their rude behavior.

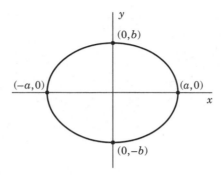

Figure 7.7 Ellipse.

ELLIPSE

This one is a circle wannabe, one that can't quite get round. Its equation is typically given by something like

$$\frac{x^2}{a^2} + \frac{y^2}{b^2} = 1$$

The number a gives the "radius" along the x-axis and the number b gives the radius along the y-axis (Figure 7.7).

We can move its center around by sticking in some constants. The same ellipse centered at (x_0, y_0) has equation

$$\frac{(x - x_0)^2}{a^2} + \frac{(y - y_0)^2}{b^2} = 1$$

Now it is possible to have ellipses that are tilted at an angle, and these will have more loopy equations. Parabolas and hyperbolas can be tilted too. But it's extremely unlikely that such eccentric goofballs will show up in your calculus class.

PARABOLA

We've already said more than enough about the parabola. This leaves that excitable curve, the hyperbola.

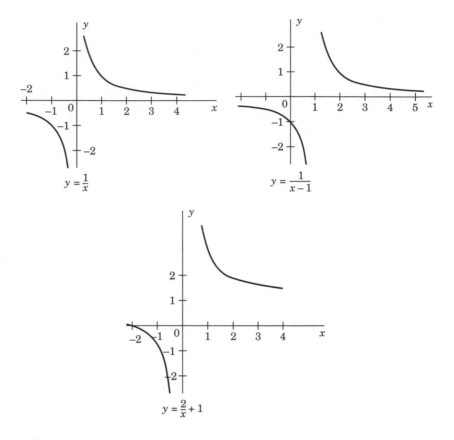

Figure 7.8 Various hyperbolas.

HYPERBOLA

The hyperbola is a curve in need of a sedative if ever there was one. Very confused, it shoots off in two different directions at the same time. The standard variety is represented by

$$y = \frac{1}{x}$$

As with its friends, we can move or stretch it by fiddling with its equation (see Figure 7.8).

We can also have hyperbolas given by equations like $y^2 - x^2 = 9$ (see Figure 7.9). Note that when $y = 0$, there is no solution for x, so the graph does not cross the x-axis.

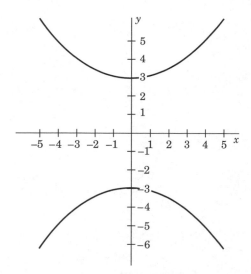

Figure 7.9 Hyperbola $y^2 - x^2 = 9$.

A GRAPHIC STORY

One professor drew an x-axis and a y-axis on the floor of his classroom and illustrated graphing by having the students walk along the graphs of functions he called out. "Applied math," he called it. One day a student showed up in class with his dog. "Get that dog out of here," yelled the professor angrily. "I don't allow dogs in my classroom!"

"But my dog knows how to graph functions," said the student.

"That's ridiculous," said the professor, "Get it out of here."

"Try it for yourself," said the student. "Give him a function to graph."

"All right then. Let's see him graph $y = 3x + 1$," said the professor.

The dog immediately trotted over to the side of the classroom and then proceeded to run along a beeline with slope 3 and y-intercept 1, astonishing the class. It then leaped up on the professor and licked his face from one ear to the other with its long, wet tongue.

Wiping his face off with his handkerchief, the infuriated professor decided to try something harder. He called out, "$x^2/4 + y^2/9 = 1$." The dog promptly ran around an ellipse passing through $(-2, 0)$, $(0, -3)$, $(2, 0)$, and $(0, 3)$. It then enthusiastically jumped up on the professor once again, thoroughly licking his face with its long, wet, slobbering tongue.

The class sat in awed amazement. The student, in triumph, called out, "Is anyone here still skeptical? Does anyone still doubt the genius of my pooch? Is there anyone else who wants to try this?"

He was answered only with silence, total silence, as the class sat stunned. Finally one student in the corner, who hadn't been heard from all semester, stuck up his hand. "I'd like to try it," he said, "but I'm not sure my tongue is long enough."

Limits: You Gotta Have Them

8.1 Basic idea

You know how people say, "You're really annoying me. You're pushing me to the limit." Did you ever wonder exactly what they meant by that? Did you ever say, "What exactly do you mean, 'pushing you to the limit'?" Did their faces turn bright red and veins stick out of their foreheads until suddenly they grasped for their chests as they keeled over? Did you then feel you knew better what they meant?

Usually, in English, the word *limit* is used to mean a boundary beyond which one cannot go. As they say in Minnesota after a successful day fishing for pike, "I've caught my limit." Any more fish than that, and the game warden confiscates your tackle box. The number of fish caught can approach the limit and perhaps reach it, but it cannot exceed it.

In mathematics, a limit is a number that a function approaches as the values of x plugged into the function approach a fixed number. Let's put your face into an example.

Suppose that you move your nose toward a fan. Think of your nose at position x and the fan at position 3, 3 feet from the origin, as shown in Figure 8.1.

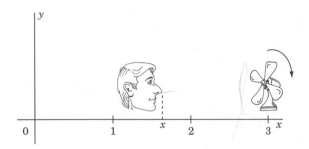

Figure 8.1 A dangerous limit to plug into.

We want to know what happens as x gets really close to 3—that is to say, what happens as your nose approaches 3, getting closer and closer, WITHOUT EVER ACTUALLY REACHING 3.

Well, what happens is that you feel a breeze that gets stronger as x gets closer to 3. We are interested in what happens to the amount of breeze as you approach 3. [We are taking $\lim_{x \to 3} b(x)$, where $b(x)$ is the breeze that you feel when your nose is at point x.]

Say you feel a breeze of 6 mph when $x = 2.9$, and the breeze increases as you move your nose toward the fan as in the following chart:

Nose position	2.9	2.99	2.999	2.9999	2.99999
Breeze	6	6.7	6.92	6.991	6.999993

It looks as if the breeze is approaching 7 mph as your nose approaches the fan. So we would say that

$$\lim_{x \to 3} b(x) = 7$$

You most certainly do not want to find out what happens when $x = 3$; you will feel a lot more than a breeze.

That's the way limits work. We want to see what happens to the values of a function $f(x)$ as x approaches a particular number a.

We can also determine the limit of a function by looking at its graph. In Figure 8.2, we see the graphs of two functions, both of which have limit equal to 7 as x approaches 3. On both graphs, as x approaches 3 from either side, the values of the function $f(x)$ approach the value 7. On graph (b), even

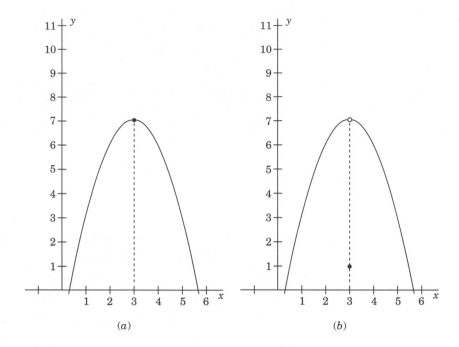

Figure 8.2 Two functions with limit equal to 7 as $x \to 3$.

though the function has a weird value of 1 at the point 3, we don't care. The limit doesn't depend on the value at 3, only on the values as we approach 3.

For most of the standard functions, nothing too interesting happens when we take a limit. For a polynomial like $x^3 - 7x^4 + 3$, the limit as x approaches a number a is just $a^3 - 7a^4 + 3$. We get the limit by plugging a in for x.

Example $$\lim_{x \to 1} x^2 - 7x^3 + 5 = 1^2 - 7(1^3) + 5 = -1$$

We just plugged in 1 for x and got the answer! If only life was always this simple. But, no big surprise, there are going to be some problem functions, some troublemakers. Look at

$$\lim_{x \to 2} \frac{1}{(x - 2)^2}$$

As x gets close to 2, $(x - 2)^2$ gets very small. But 1 divided by a tiny number is a very large number, and we mean LARGE. So, as x approaches 2, $\frac{1}{(x - 2)^2}$

approaches ∞. Since ∞ is not a real number, we say that THE LIMIT DOES NOT EXIST. It's just not there, it's a nonentity, no answer, dead on arrival. Some people use the notation DNE for "does not exist." Use this only if your instructor does.

So we can have limits that are so big (bigger than any real number) that the limit doesn't exist. We can also have limits that don't exist for other reasons. For instance, take a look at

$$\lim_{x \to 0} \sin\left(\frac{1}{x}\right)$$

As x gets smaller and smaller, $1/x$ gets bigger and bigger. But $\sin(1/x)$ will always be between -1 and 1, since the sine of any number is between -1 and 1. As $1/x$ gets bigger and bigger, $\sin(1/x)$ just oscillates between -1 and 1, faster and faster. It goes crazy. In particular, it is not getting closer and closer to any one number. It's not zeroing in on any one value. So it has no limit.

Think of having a limit as falling in love with one number. As the crucial moment of commitment approaches, we fall deeper and deeper (approach) in love with this one number. But $\sin(1/x)$ can't commit. This function wants to play the field with a whole interval's worth of numbers. So in the end, no limit exists, and $\sin(1/x)$ never ties the knot. A perpetual swinger, it oscillates forever between $+1$ and -1, bouncing from one Club Med to the next.

Trick   *Hard limits can be easy at other points.*

$$\lim_{x \to 2/\pi} \sin\left(\frac{1}{x}\right)$$

This limit equals

$$\sin\left(\frac{1}{2/\pi}\right) = \sin\left(\frac{\pi}{2}\right) = 1$$

The limit of $\sin(1/x)$ has problems only as $x \to 0$, not when x approaches other values.

8.2 General procedure for taking a limit

Say we want to take $\lim_{x \to b} f(x)$ for some function $f(x)$.

The first step is always to plug b into the function. If we get a number (that doesn't have 0 in the denominator or a negative number inside a square root), and if the function is not one of those weird multiple personality guys that changes its definition at the point b, then that number $f(b)$ is the limit.

Just plugging in always works for a polynomial. In fact, it usually works for almost any function as long as the point you're plugging in doesn't make you divide by zero $\left(\text{like } \lim\limits_{x \to 0} \dfrac{1}{x}\right)$. Plugging in is called "direct substitution," or on casual occasions, "plug and chug."

Example $\quad \lim\limits_{x \to 2} x^4 - 6x^3 + 2 = 2^4 - 6(2^3) + 2 = 16 - 48 + 2 = -30$

Example
$$\lim_{x \to \pi/4} \sin x = \sin\left(\frac{\pi}{4}\right) = \frac{\sqrt{2}}{2}$$

In fact, it often works for a fraction of polynomials (a *rational function*).

Example
$$\lim_{x \to 1} \frac{x^2 - 4x + 2}{3x^2 + 6} = \frac{1^2 - 4(1) + 2}{3(1^2) + 6} = -\frac{1}{9}$$

Example
$$\lim_{x \to 3} \frac{x^2 - 9}{x - 7} = \frac{0}{-4} = 0$$

There is no problem with having a zero in the numerator. We do get worried when there is one in the denominator.

Example
$$\lim_{x \to 1} \frac{x - 1}{x^2 - 1}$$

If we just plug in, we get $\frac{0}{0}$. Don't simplify this to equal 1! It is *not* true that $\frac{0}{0}$ will just give us 1! And don't give up either.

Instead, we can simplify this fraction. Anytime we get $\frac{0}{0}$, we should always try to simplify, hoping to get rid of one of the zeros.

$$\lim_{x \to 1} \frac{x - 1}{x^2 - 1} = \lim_{x \to 1} \frac{x - 1}{(x - 1)(x + 1)} = \lim_{x \to 1} \frac{1}{x + 1} = \frac{1}{2}$$

Ta-da!

This function $\dfrac{x-1}{x^2-1}$ isn't even defined at $x = 1$. Elsewhere it looks just like $\dfrac{1}{x+1}$. But at $x = 1$ it has a hole, a vacuum in its life, a lack of meaning, a void that cannot be filled. This is not a cause for alarm. When evaluating this limit, we only care about the values at points *near* 1, and *not* the actual value at 1.

Example
$$\lim_{x \to 0} \frac{1 + x^2}{x^2}$$

If we just try to plug in here, we get a 1 in the numerator and a 0 in the denominator, namely, $\frac{1}{0}$. That means that the limit DOES NOT EXIST. (The denominator is getting smaller and smaller, and when we divide 1 by a positive number getting smaller and smaller, the result is a number that gets bigger and bigger and bigger and bigger, approaching $+\infty$.)

8.3 One-sided limits

We haven't been too careful about just how x is sneaking up on a value. In the case of

$$\lim_{x \to 2} \frac{1}{x-2}$$

x could be larger than 2, and edging toward 2 from the right, or x could be smaller than 2 and sliding up to it from the left. In both these cases, x could be in serious trouble if prosecuted under the stalking statutes. Also, in both cases the limit will not exist, but we may still want to distinguish between them.

If we want to only let x approach 2 from the right, we write

$$\lim_{x \to 2^+} \frac{1}{x-2}$$

Then $x - 2$ will be a very small POSITIVE number and

$$\frac{1}{x-2}$$

will be a very large positive number. So

$$\lim_{x \to 2^+} \frac{1}{x-2} = +\infty$$

and the limit Does Not Exist.

On the other hand, if we only let x approach 2 from the left, we write

$$\lim_{x \to 2^-} \frac{1}{x-2}$$

Since x is less than 2, $x-2$ is a small but NEGATIVE number. So $\frac{1}{x-2}$ is a large but negative number and

$$\lim_{x \to 2^-} \frac{1}{x-2} = -\infty$$

and again the limit Does Not Exist.

This concept is illustrated when two dinosaurs hunt together, as in *Jurassic Park*. The point a is played by a small mammal, say a goat. A single Tyrannosaurus x sneaking up on the goat from the left is described by

$$\lim_{x \to a^-} L(x)$$

where $L(x)$ is the lunch function. A second Tyrannosaurus x sneaking up from the right is described by

$$\lim_{x \to a^+} L(x)$$

If both dinosaurs hunt together, each stalking the goat from one side, they share a nice lunch when they reach a. We conclude that

$$\lim_{x \to a} L(x)$$

exists and equals rack of goat, very rare.

> When we say that a limit exists, we mean that both the left and right limits exist and that they are equal.

P.S. The goat never had a chance.

8.4 Limits of weird functions

Making life more complicated, mathematicians invented functions that are defined not by a single formula but by several. Large curly braces are a tip-off that you're dealing with one of these multiple personality functions. The absolute value function is one such guy. Another example is

$$f(x) = \begin{cases} 1/x & \text{if } x < 3 \\ x^2 - 12 & \text{if } x \geq 3 \end{cases}$$

Huh? We know this doesn't look like most of the functions we have been dealing with up to now, but it's perfectly valid. We could graph it, and it would look like Figure 8.3.

Let's look at some limits for this function.

1. What about $\lim_{x \to 4} f(x)$? For $x \geq 3$, $f(x) = x^2 - 12$, so

$$\lim_{x \to 4} f(x) = \lim_{x \to 4} (x^2 - 12) = 4.$$

2. $\lim_{x \to 2} f(x)$? For $x < 3$, $f(x) = 1/x$, so $\lim_{x \to 2} f(x) = \lim_{x \to 2} (1/x) = \tfrac{1}{2}$.

3. $\lim_{x \to 3^+} f(x)$? For $x \geq 3$, $f(x) = x^2 - 12$, so $\lim_{x \to 3^+} f(x) = -3$.

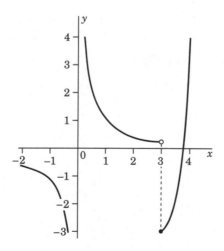

Figure 8.3 Function that changes its personality.

4. $\lim\limits_{x \to 3^-} f(x)$? For $x < 3$, $f(x) = 1/x$, so $\lim\limits_{x \to 3^-} f(x) = 1/3$. When we do this calculation, the value of $f(x)$ at the actual point $x = 3$ is IRRELEVANT. Don't care. Forget about it. It doesn't affect our calculation at all.

5. $\lim\limits_{x \to 3} f(x)$? Well, for $x < 3$, the limit would be $1/3$, and for $x \geq 3$ the limit would be -3, so the left and right limits at 3 can't come to an agreement. Like two siblings punching each other in the back seat of the car, their inability to get along ultimately causes the absolute worst-case scenario, namely, one of them has to sit in front with the parents, or in our case, the full limit does not exist.

Strange but True Department   You may not want to deal with this, but it's an important fact in the world of limits that

$$\lim_{x \to 0} \frac{\sin x}{x} = 1$$

Sort of strange, an ugly function like $\dfrac{\sin x}{x}$ having a pretty limit like 1. In the greater scheme of things, what does this mean? It means that as x gets close to 0 (that is to say, very very small), $\sin x$ behaves just like x. Their ratio $\dfrac{\sin x}{x}$ goes to 1. The best way to visualize this is to graph the two functions $y = x$ and $y = \sin x$ as in Figure 8.4. As x gets close to 0, the functions become more and more alike.

The function $y = x$ gives a pretty good approximation for the function $y = \sin x$, for very small values of x. So we should not be surprised that the ratio $\dfrac{\sin x}{x}$ is getting close to 1.

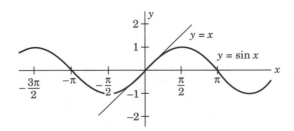

Figure 8.4 $y = x$ and $y = \sin x$ look pretty similar near $x = 0$.

Important Technical Note This is true only if x is measured in radians, rather than in degrees. When you see the function $\sin x$ in calculus, it is always assumed that x is given in radians.

Hard Exam Question Find $\lim_{x \to 0} \frac{\sin 5x}{x}$.

Well, rewrite this as

$$\lim_{x \to 0} \frac{5 \sin 5x}{5x}$$

by multiplying the fraction by $\frac{5}{5}$. When $x \to 0$ then $5x \to 0$ too, and vice versa, so this is the same as

$$\lim_{5x \to 0} \frac{5 \sin 5x}{5x}$$

Writing $z = 5x$, we can write this as

$$\lim_{z \to 0} \frac{5 \sin z}{z} = 5 \lim_{z \to 0} \frac{\sin z}{z} = 5(1) = 5$$

Good Question to Ask Instructor What exactly do you mean by *infinity*? Ask this when the instructor says something like "the limit as x goes to infinity."

Wrong answer: The biggest number, a godzillion, roughly 10^{100}.
Right answer: A shorthand way of expressing that x is taking on arbitrarily large values.

Another Good Question What does "undefined" mean? You could ask this when the instructor says something like "this limit is undefined."

Wrong answer: Lacks muscle tone. Flabby.
Right answer: Not equal to any single number.

Common Mistake When you're trying to figure out $\lim_{x \to a} f(x)$ and you plug in the value a in for x, sometimes you will get a fraction that looks like $\frac{0}{0}$. In this case you know NOTHING about the limit. A common mistake is to say that the limit does not exist. An even worse mistake is to cancel the zeroes and say that the limit equals one. In truth, the $\frac{0}{0}$ tells you that more work must be done to determine the limit, usually some cancellation. Keep in mind the example $\lim_{x \to 0} \frac{3x}{x}$. The answer is 3, and easy IF you do some cancellation. Generally, if you get something like $\frac{3}{0}$ you know the limit is infinite, if you get something like $\frac{0}{3}$ then the limit is 0, and if you get $\frac{0}{0}$ you know nothing, and more work is required.

 8.5 Calculators and limits

Since we know that a limit is just a value approached by a function $f(x)$ as x approaches a particular value a, we should be able to get a good idea of what a limit is by actually evaluating $f(x)$ on a computer or calculator as x approaches a to see what it is approaching.

For instance, one can see that $\lim\limits_{x \to 0} \dfrac{\sin x}{x}$ is likely to be 1 by using a calculator and just plugging into $\dfrac{\sin x}{x}$ values for x that approach 0, such as $1, .1, .01, \ldots$. The values that the calculator generates will be approaching 1. Here is a short program that does eight such calculations.

Sample Program for Approximately Computing a Limit

1 $A = 0$

2 $f(x) = \dfrac{\sin x}{x}$

3 FOR $k = 1$ TO 8

4 $A = f(10^{1-k})$

5 PRINT A

6 NEXT k

This program plugs values of x getting closer and closer to 0 into the function $\dfrac{\sin x}{x}$ and prints out each result. As you will see, the values are quickly approaching 1. In fact, for any limit, one could try such a program to try to see where it was headed.

Continuity, or Why You Shouldn't Ski Down Discontinuous Slopes

9.1 The idea

In life, when people say, "Hey, let's get some continuity here," it means that they don't want dramatic changes every four minutes. They do not want the entire staff to be fired and replaced by a bunch of tenderfoots. They would like the normal way of doing things to continue uninterrupted. That's the key word, "uninterrupted." A function is continuous if it continues on its way uninterrupted, as for instance occurs in the graphs in Figure 9.1.

These graphs can be drawn without ever lifting your pen from the page. You can draw them in one continuous motion. However, neither graph in Figure 9.2 can be drawn in one continuous motion. Each of them has a problem at the point $x = 2$. In each of these cases we say that the function is *discontinuous* at $x = 2$.

9.2 Three conditions for continuity

It's time for the official definition of "continuous." You know this is serious business because it has three parts.

Figure 9.1 Continuous functions.

Definition *f(x) is continuous at x = a*

1. *f(a) is defined*

2. $\lim_{x \to a} f(x)$ *exists*

3. $\lim_{x \to a} f(x) = f(a)$

Let's see what this means.

* Condition 1 means that the point we are considering is in the domain of the function. We can't talk about whether the function $f(x) = \sqrt{x}$ is continuous at $x = -4$, since it isn't even defined on negative numbers.

* Condition 2 means that the function is approaching a particular value as x gets closer and closer to a from either side of a.

* Condition 3 says that the number that $f(x)$ is supposed to be getting closer to is the value of the function at a.

Let's see how each of these conditions can fail. In Figure 9.3, the function is defined at 2 and the limit exists at 2, but they are not equal to each other. Condition 3 fails. This causes there to be a momentary jump down and then back up in the graph of the function.

In Figure 9.4, the function is defined at 2; however, the limit does not exist, since the limit we get as we approach 2 from the right is different from the limit we get as we approach 2 from the left. Condition 2 fails.

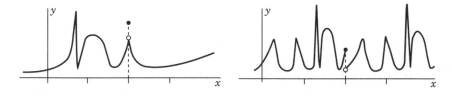

Figure 9.2 Discontinuous function with a puncture, and one with a step.

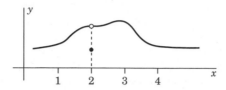

Figure 9.3 Discontinuous function with a puncture.

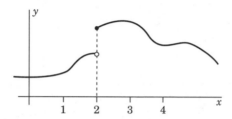

Figure 9.4 Discontinuous function with a step.

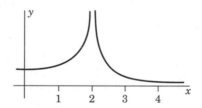

Figure 9.5 Discontinuous function with an infinite limit.

Finally, in Figure 9.5, the roof blows off. Not only is the function undefined at $x = 2$, but the limit doesn't exist either. This time we see the graph heading off the top of the page. All three conditions fail.

If there are no breaks anywhere in the graph, the function is continuous for all values of x—it is continuous everywhere. So what functions have this skier-friendly attribute?

Polynomials like x^5 or $2x^3 + 5x^2 - 3x + 8$ are continuous everywhere. So are $\sin x$ and $\cos x$. The function $|x|$ is continuous everywhere, even at the kink, since it can still be drawn without lifting pen from page. Fractions of polynomials, $p(x)/q(x)$ where p and q are polynomials, are continuous for all values where $q(x)$ is not equal to 0, which is all points where the functions have a value. (These are sometimes called *rational functions*.) For instance, $(x^2 - 1)/(x + 3)$ is continuous except at $x = -3$. The functions $\tan x$, $\sec x$, and $\csc x$ are continuous everywhere they are defined, which is *not* everywhere.

So $\tan x = \dfrac{\sin x}{\cos x}$ is continuous except where $\cos x = 0$, namely, at the points $x = \pi/2, x = 3\pi/2, \ldots$.

In a nutshell, *all* the standard functions that you deal with every day are continuous everywhere they are defined. So life is looking easy and the answer to the question "Where is function blah continuous?" is almost always "Everywhere blah is defined."

But there are exceptions to this rule when we throw in the multiple personality functions. These have different definitions at different values of x. We need to examine each of the points where they fit together, one point at a time.

Example   *Determine where the following function is discontinuous.*

$$f(x) = \begin{cases} 0 & \text{if } x \leq -1 \\ x + 1 & \text{if } -1 < x \leq 0 \\ x^2 & \text{if } 0 < x < 1 \\ 1 & \text{if } x = 1 \\ \dfrac{1}{2 - x} & \text{if } 1 < x \end{cases}$$

This function is defined everywhere except at $x = 2$, where $\dfrac{1}{2 - x}$ has a gap. The functions that are used in its definition, 0, $x + 1$, x^2, and $\dfrac{1}{2 - x}$, are continuous everywhere they are defined, so we need only check the continuity of $f(x)$ at the points where the functions are pieced together, namely, at $x = -1, 0,$ and 1.

At $x = -1$,

$$\lim_{x \to -1^-} f(x) = \lim_{x \to -1^-} 0 = 0$$

and

$$\lim_{x \to -1^+} f(x) = \lim_{x \to -1^+} x + 1 = -1 + 1 = 0$$

The left and right limits are equal. So the full limit exists. In addition, the value of the function at $x = -1$ is 0, so

$$\lim_{x \to -1} f(x) = 0 = f(-1)$$

and the function satisfies all three conditions needed for it to be declared continuous at $x = -1$.

At $x = 0$,

$$\lim_{x \to 0^-} f(x) = \lim_{x \to 0^-} (x + 1) = 1$$

and

$$\lim_{x \to 0^+} f(x) = \lim_{x \to 0^+} x^2 = 0$$

Both one-sided limits exist, but the left and right limits at 0 ARE NOT EQUAL, so the full limit does not exist. So $f(x)$ is not continuous at $x = 0$.

At $x = 1$,

$$\lim_{x \to 1^-} f(x) = \lim_{x \to 1^-} x^2 = 1$$

and

$$\lim_{x \to 1^+} f(x) = \lim_{x \to 1^+} \frac{1}{2 - x} = 1$$

The left and right limits are equal. So the full limit exists. Since the value of the function at $x = 1$ is 1,

$$\lim_{x \to 1} f(x) = 1 = f(1)$$

All three conditions are fulfilled, and the function is continuous at $x = 1$.

End result? Ultimate bottom line? The function is continuous at all values of x except $x = 0$ and $x = 2$. Figure 9.6 gives a picture of the graph.

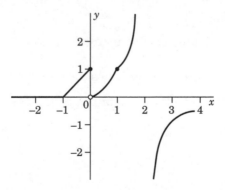

Figure 9.6 Discontinuous function with a step and a missing point, where it goes infinite.

Example *Determine where the function $f(x) = \frac{|x-2|}{x-2}$ is continuous.*

The first thing we notice about this function is that there is a zero in the denominator when $x = 2$. Bingo! That's all we need to know that the function is *not* continuous at $x = 2$. It isn't even defined there! The value $x = 2$ is not in the domain of the function. What about other points?

This function doesn't look like a piecemeal function, but in fact, it is one in disguise. Let's use the definition of the absolute value function to write this function out in a more revealing form.

$$\frac{|x-2|}{x-2} = \begin{cases} \dfrac{x-2}{x-2} & \text{if } x - 2 > 0 \\ \dfrac{-(x-2)}{x-2} & \text{if } x - 2 < 0 \end{cases}$$

Notice that the function is defined when $x - 2 > 0$ and when $x - 2 < 0$, but *not* when $x - 2 = 0$. The point $x = 2$ is not in the domain, so the function is not defined at that point.

We can simplify the expressions to get

$$\frac{|x-2|}{x-2} = \begin{cases} 1 & \text{if } x - 2 > 0 \\ -1 & \text{if } x - 2 < 0 \end{cases}$$

Another way to write $x - 2 > 0$ is to write $x > 2$ instead. Another way to write $x - 2 < 0$ is to write $x < 2$ instead. That is to say, our fancy function $f(x)$ is just given by

$$\frac{|x-2|}{x-2} = \begin{cases} 1 & \text{if } x > 2 \\ -1 & \text{if } x < 2 \end{cases}$$

Now, 1 and -1 are continuous functions. So $\dfrac{|x-2|}{x-2}$ is *not* continuous at $x = 2$, and it *is* continuous everywhere else. Notice that $\lim\limits_{x \to 2} \dfrac{|x-2|}{x-2}$ doesn't exist either, since the left and right limits at 2 can't come to agreement.

What Is the Derivative? Change Is Good

Okay, here we are at the essential concept of calculus, the single most important moment in your nascent calculus career. What's the derivative? What's all the fuss about? Why does everybody make such a big stink about this one simple idea?

Well, the derivative is really pretty simple. It can be summed up in one word: "slope."

Example (Gotta Goat) Suppose that you are going to walk up a hill, carrying a tranquilized goat. Let's give the base of the hill the coordinates $x = 0$ and $y = 0$. Now, as you walk up the hill, both your x and y coordinates are going to change. In fact, each of them will increase. Let's take $h(x)$ to be the height of the hill above the point x. So the graph of the function $h(x)$, which is a plot of the points satisfying the equation $y = h(x)$, is simply a silhouette of the hill, as shown in Figure 10.1.

For you, the most interesting aspect of the hill at any given point is how steep it is, since the steeper it is, the more difficult it is going to be to carry the comatose goat up that part of the hill. The derivative of the function $h(x)$ is exactly the steepness of the hill at point x. We denote it by $h'(x)$.

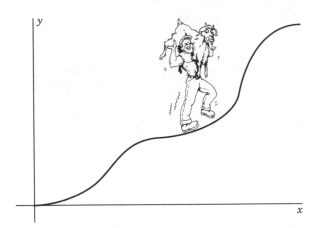

Figure 10.1 Hill with height given by the function $h(x)$.

For instance, suppose $h'(10) = \frac{1}{6}$. After having reached a point that is 10 feet out in the x direction, you are at a place on the hill that has steepness $\frac{1}{6}$. This means that you must go up about 2 inches vertically for every foot that you travel horizontally, not too steep.

On the other hand, suppose that $h'(20) = 5$. Then, after having traveled 20 feet in the x direction, you find yourself at a point on the hill that is much steeper. Here, you must go up about 5 feet vertically for every 1 foot horizontally. For that you need mountaineering equipment and a winch for the goat.

Finally, what if $h'(30) = -2$? That would mean that when $x = 30$, you are heading in the negative vertical direction 2 feet for every foot you go in the horizontal direction. In other words, you are now headed downhill. You could just roll the goat down this stretch (see Figure 10.2).

Of course, derivatives are useful in more general situations than getting drugged goats up hills. They also work for drugged sheep, drugged prairie dogs, and even small drugged water buffalo. And they even work for functions that measure things other than the elevation of a goat.

Example (Owls) Suppose you're in the business of selling plastic spotted owls, and a team of plastic lawn ornament specialists (called MBAs) has determined a function that gives the amount of profit you will earn as a function of the price that you charge per owl. From this, you want to determine the right price to charge per owl to maximize your profit.

You need to know where the profit function peaks. But if you draw the graph of your function, the peak occurs when the function is momentarily neither increasing nor decreasing (Figure 10.3). For just a second, its rate of change is

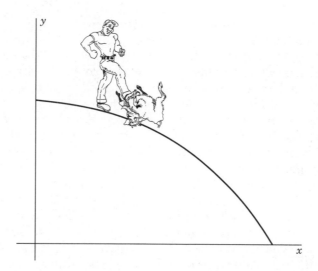

Figure 10.2 Negative slope, and a bad day for a goat.

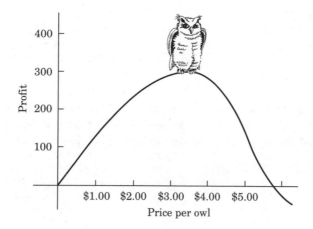

Figure 10.3 Owl profit function.

0. Putting that into high falutin' math terms, "the derivative is 0 at the peak." Simple enough. We can then use that fact to find the peak and determine the price at which to sell these scuff-resistant owls.

Example (At the races)   In our third example, we assume that you are going to be in a drag race. Here, we mean a race with a bunch of cars, not a group of guys in tutus and running shoes. There you are, sitting at the starting line in your super-charged gasoline-guzzling speedmobile, gunning the engine. Once the race begins, $f(t)$ will tell you your distance from the starting line at time t in the race, where $t = 0$ is when the gun goes off to start the race.

In a race like this, you are very interested in your speed. That's why the cars have speedometers. But speed is just the rate of change of position, right? As in 110 miles per hour. That is a certain distance per fixed time. If $f(t)$ is the position function, then the derivative $f'(t)$ is the rate of change of that function, which is exactly the speed. As the race progresses, your speed will change. It starts out at 0 mph, and then it increases from 0 to your car's maximum speed of 130 mph. And when you pass the finish line, the tail chutes pop out and you decelerate again. So over the course of the race, your derivative went from 0 up to 130 and then back down to 0. The speedometer is just telling you what the derivative of your position is at any given time. It could just as easily be called the derivativeomometer, except that's a bit of a mouthful, and would clutter up the car parts catalogs.

What if, before the race started, you put the car into reverse by mistake? The green light came up, signaling the beginning of the race. You put the pedal to the metal and found yourself rocketing backward. Of course, you would never know it from looking at your speedometer. Those stupid things don't register negative numbers. But really you are going -130 mph relative to the direction you were expecting. You know you are going in reverse because in your rearview mirror you see the terror-stricken faces of the pit crew growing in size at an incredible rate. So you would say that the derivative of your position function is -130.

A Quick Recap So far, the point of this whole section has been to get across the idea that the derivative $f'(x)$ of a function $y = f(x)$ measures the rate of change of the function. A large positive derivative says the function

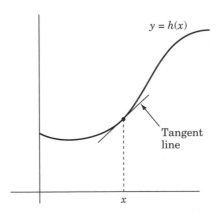

Figure 10.4 Steepness is measured by the slope of the tangent line at $(x, h(x))$.

is increasing fast. A small positive derivative says the function is increasing slowly. A negative derivative says that the function is actually decreasing. A derivative of 0 says the function is dead level, at least for an instant—put it down as undecided, going nowhere.

So great, that's the derivative. But of course, we are left with one very basic question. How do you actually calculate it? How do you measure the steepness of a hill, for instance? Well, another word for steepness is slope. So we would like to measure the slope of the hill at any point. But the hill is described by a curve and we have no means to measure the slope of a curve. The only thing we really know how to measure the slope of is a LINE.

So what we'll do is take a line that has the same steepness as the hill, measure its slope, and that will be the derivative.

This line that just touches the hill at the point $(x, h(x))$ and that has the same steepness as the hill there is called a *tangent line* (see Figure 10.4). Next we'll see how to actually compute the slope of the tangent line.

Limit Definition of the Derivative: Finding Derivatives the Hard Way

11.1 Defining the derivative

This is where we explain the official definition of the derivative. No more of this vague talk about it measuring "how fast the function is changing" or about slope analogies. No, we are going to be explicit here, completely explicit.

First of all, what would we like the derivative to measure? Well, actually, we do want it to measure how fast a function is changing. But what does that mean? Let's look at the graphs for two functions, $y = g(x)$ and $y = f(x)$, shown in Figure 11.1.

Notice that at the point x marked on the x-axis, the graph of $g(x)$ is going up much faster than the graph of $f(x)$. If these graphs represented our profit function over time, we would much prefer $g(x)$ to $f(x)$. The reason is that $g(x)$ is a lot steeper, meaning our profit is going up a lot faster.

Of course, we would like to know exactly how much faster it's going up, so as to be able to impress our relatives with our business acumen when we see them over Thanksgiving. We want to be able to measure precisely how steep that graph is at the point on the graph above the point x.

The steepness of a hill is also called the slope of the hill. So what we really want to measure is the slope of the graph at the point on the graph above x.

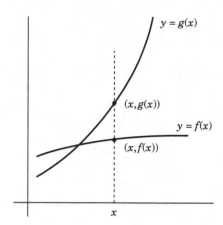

Figure 11.1 How steep are the graphs at the given point?

But just what is the slope of a curve? We know how to make sense of the slope of a straight line. So, instead of measuring the slope of the curve, we will take a line through the point $(x, g(x))$ that is tilted with the same steepness as the curve and then measure the slope of that line. We call that the slope of the curve.

The line that we are interested in, the one that approximates the curve near the point $(x, g(x))$, is called the *tangent line* to the curve (see Figure 11.2).

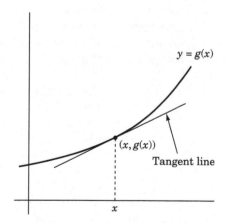

Figure 11.2 The tangent line has the same slope as the graph at $(x, g(x))$.

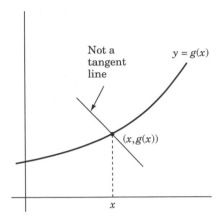

Figure 11.3 A line that intersects the curve at just one point doesn't have to be tangent.

So the derivative of the function $y = g(x)$ at x is equal to the slope of the tangent line to the curve at the point $(x, g(x))$.

Of course, a good question is "How do you even know which line is the tangent line at the point $(x, g(x))$?" Is it the line that intersects the curve just at the point? No, Figure 11.3 shows a line that does that but it certainly is not the tangent line.

So we need to be careful about the way we even define a tangent line. We want it to be the line that just barely skims by the curve at a point, heading in essentially the same direction as the curve.

How about this idea? Let's take another point on the curve with x-coordinate just a teensy distance h to the right of x. Its coordinates are $(x + h, g(x + h))$. (Keep in mind that its y coordinate is just the function evaluated at its x coordinate. It better be, if we want the point to be on the graph of the function.)

Now, take the line that passes through this point and the point $(x, g(x))$. That isn't the tangent line, but it's not so far away from the tangent line. We will call that line a *secant line*, and denote it by S. (See Figure 11.4.)

Let's calculate the slope of the secant line, called slope(S). We get

$$\text{Slope}(S) = \frac{y_1 - y_0}{x_1 - x_0} = \frac{g(x + h) - g(x)}{(x + h) - x} = \frac{g(x + h) - g(x)}{h}$$

Notice as we move the second point toward the first along the graph of $g(x)$ (by letting h shrink to 0), the secant line S swivels toward the tangent line T. It is "approaching" the tangent line (see Figure 11.5).

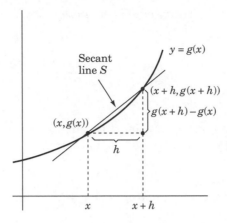

Figure 11.4 A secant line approximates the slope at $(x, g(x))$.

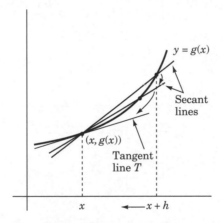

Figure 11.5 Secant lines getting close to the tangent line.

So as the second point moves toward the first, the slope of the secant line, slope(S), approaches the slope of the tangent line, slope(T), which is what we are after. Or, slope(S) $\rightarrow$ slope(T) as $h \rightarrow 0$. We write this fact as

$$\text{Slope}(T) = \lim_{h \to 0} \frac{g(x + h) - g(x)}{h}$$

Since we want the slope of the tangent line at x to be the derivative, we state the following.

Definition   *The derivative of $g(x)$ evaluated at x is written $g'(x)$ and is given by*

$$g'(x) = \lim_{h \to 0} \frac{g(x + h) - g(x)}{h}$$

This is just a limit, no fancier than any of the other limits we have already discussed. We can actually use this to compute a real live derivative.

Example   *Find the derivative of $f(x) = x^2$.*

We want to figure out

$$f'(x) = \lim_{h \to 0} \frac{f(x + h) - f(x)}{h}$$

We know that $f(x) = x^2$, and by replacing each x in x^2 with an $x + h$, we have that $f(x + h) = (x + h)^2$.

So

$$f'(x) = \lim_{h \to 0} \frac{(x + h)^2 - x^2}{h}$$

$$= \lim_{h \to 0} \frac{x^2 + 2xh + h^2 - x^2}{h}$$

$$= \lim_{h \to 0} \frac{2xh + h^2}{h}$$

$$= \lim_{h \to 0} \frac{h(2x + h)}{h}$$

$$= \lim_{h \to 0} (2x + h)$$

$$= 2x$$

(Since the $2x$ doesn't involve h, the $2x$ is left alone as we take the limit and the h becomes 0.)

So if

$$f(x) = x^2$$

then

$$f'(x) = 2x$$

Literally, this means that if you give us a value for x, we can tell you the slope of the tangent line to $f(x) = x^2$ for that x value. For instance, you say $x = 1$; then we say

$$f'(1) = 2(1) = 2$$

Looking at the graph in Figure 11.6, and estimating the slope of the tangent line at $x = 1$, this is believable. It looks to be about 2.

You say $x = -1$, and we say

$$f'(-1) = 2(-1) = -2$$

Again this should be believable based on the picture. That line definitely has a negative slope of -2.

Finally, you say $x = 0$, and we say

$$f'(0) = 2(0) = 0$$

Therefore, the tangent line at $x = 0$ should have slope 0, and therefore be horizontal. Lo and behold, it is. Pretty amazing, huh?

Let's do another famous example.

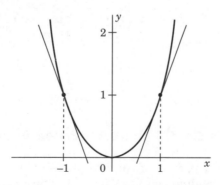

Figure 11.6 Tangent lines to $y = x^2$ at $x = 1$ and $x = -1$.

Example   *Find $f'(x)$ if $f(x) = \sqrt{x}$.*

Then

$$f'(x) = \lim_{h \to 0} \frac{\sqrt{x+h} - \sqrt{x}}{h}$$

$$= \lim_{h \to 0} \left(\frac{\sqrt{x+h} - \sqrt{x}}{h} \right) \left(\frac{\sqrt{x+h} + \sqrt{x}}{\sqrt{x+h} + \sqrt{x}} \right)$$

$$= \lim_{h \to 0} \frac{(x+h) - x}{h(\sqrt{x+h} + \sqrt{x})}$$

$$= \lim_{h \to 0} \frac{h}{h(\sqrt{x+h} + \sqrt{x})}$$

$$= \lim_{h \to 0} \frac{1}{\sqrt{x+h} + \sqrt{x}}$$

$$= \frac{1}{2\sqrt{x}}$$

If you know that little algebra trick of multiplying by $\dfrac{\sqrt{x+h} + \sqrt{x}}{\sqrt{x+h} + \sqrt{x}}$, you're on easy street in this calculation. Otherwise you're up Dead End Alley.

   WARNING: Sometimes $\displaystyle\lim_{h \to 0} \frac{g(x+h) - g(x)}{h}$ doesn't exist. Then we say that the derivative $g'(x)$ is undefined.

11.2 Other forms for the limit definition of the derivative

Just as there is a variety of colorful phrases to express "Hey, your bumper has inexplicably collided with my bumper, and I believe you are responsible," there are also several variations on the basic definition of the derivative.

But they are essentially the same. Use whichever one you feel closest to. For instance, often people replace the h we used above with Δx to get

$$f'(x) = \lim_{\Delta x \to 0} \frac{f(x + \Delta x) - f(x)}{\Delta x}$$

For those not in fraternities or sororities, Δ is the Greek capital letter delta. Unfortunately, Isaac Newton was denied the experience of the brotherhood of fraternities due to a genetic inability to master secret handshakes. He must have thought it an ironic twist to make Δ, the very symbol of his greatest disappointment, the cornerstone of his greatest achievement. The symbol Δx is used to represent a small change in x.

We could make more substantial changes in the original definition. For instance, first, let's replace x by a. Why? Why not? So now we have

$$f'(a) = \lim_{h \to 0} \frac{f(a + h) - f(a)}{h}$$

Now we let a new $x = a + h$. Then as h goes to 0, x approaches a. So we would write

$$f'(a) = \lim_{x \to a} \frac{f(x) - f(a)}{x - a}$$

Looks pretty different, right? But it's just the same thing, recycled. Like when they make old soda bottles into carpeting.

When do you use the limit definition to compute a derivative? Once in a million years, but this section is that once. For all but the most unusual functions, we will develop a set of straightforward rules for finding the derivative in the next section. These rules will be much simpler to use than directly applying the limit definition of the derivative.

The derivative of a function $f(x)$ is written in a variety of ways, as

$$f'(x)$$

or

$$\frac{df}{dx}$$

or that favorite of those wacky kids, the physicists,

$$\dot{f}$$

Often we let $y = f(x)$, in which case we have three more options for writing the derivative:

$$y'$$

or

$$\frac{dy}{dx}$$

or

$$\dot{y}$$

Derivatives: How to Find Them the Easy Way

Here we are at a core topic. This is what you'll remember about calculus when you're in the nursing home sipping your dinner through a straw. Fortunately, the basic mechanics of calculating derivatives are not that hard to master.

12.1 Basic rules of differentiation

What can we say? Know these backward and forward. They are to calculus what "Don't go through a red light" and "Don't run over a pedestrian" are to driving.

12.2 Power rule

Here is one of the most basic rules of calculus, perhaps the single rule that more people think of when someone says, "What do you remember about calculus?" Besides, it has that great name, the **power rule.**

$$\frac{d}{dx}(x^n) = nx^{n-1}$$

Now what does that mean? Well, n can be any number. So if we let n be 3, we have

$$\frac{d}{dx}(x^3) = 3x^2$$

If we let n be 5, we have

$$\frac{d}{dx}(x^5) = 5x^4$$

If we let n be 1, we have

$$\frac{d}{dx}(x^1) = 1x^0 = 1$$

This is worth remembering separately, so let's repeat it so it sticks:

$$\frac{d}{dx}(x) = 1$$

We could also let n be a negative number. For instance, if $n = -2$, we have

$$\frac{d}{dx}(x^{-2}) = -2x^{-3}$$

Notice that we can now take the derivative of $1/x$:

$$\frac{d}{dx}\left(\frac{1}{x}\right) = \frac{d}{dx}(x^{-1}) = -1x^{-2} = \frac{-1}{x^2}$$

In fact, n can even be a fraction. So, for instance, we can take the derivative of $\sqrt{x}$, since

$$\frac{d}{dx}\left(\sqrt{x}\right) = \frac{d}{dx}(x^{1/2}) = \frac{1}{2}x^{-1/2}$$

The fact that n needn't be a whole number usually doesn't get mentioned in this part of a course on calculus, but we'll clue you in a little early. That little n can even be a number like π or $\sqrt{2}$. Then

$$\frac{d}{dx}(x^\pi) = \pi x^{\pi - 1}$$

We know it's tempting to try to simplify the exponent $\pi - 1$, but it's best to leave it the way it is. Just think of it as a number with value about 2.1416.

Good Question   Is it possible to prove the power rule for whole numbers using just induction and the product rule?

Of course, there is a little risk here. This question makes you sound exceptionally intelligent and will momentarily lift you onto a godlike plateau miles above the rest of the class in the mind of the professor. If the professor swallows the bait, and uses induction and the product rule to prove the power rule, you're set for life, or at least for the rest of the semester. However, if the professor doesn't bite but instead says, "How do you mean?" and you respond, "Um, ah, oh I don't know, I just made it up," it could look very bad.

12.3 Product rule

Okay, here is a key rule, a rule without which calculus would just be a case of goofing around with a pile of very simple functions. Suppose we have a function h that is obtained by multiplying together two functions f and g. So $h(x) = f(x)g(x)$. And silly us, we want to take the derivative of h. It's extremely tempting to say $\frac{d}{dx}(fg) = f'g'$, but unfortunately that would be WRONG!

How do we know it's wrong? Well, we didn't waste the best years of our lives studying math for nothing. Besides, if we apply it when $f(x) = x$ and $g(x) = x$, we get $\frac{d}{dx}(x)(x) = (1)(1) = 1$. WHICH IS WRONG! We already know that

$$\frac{d}{dx}(x)(x) = \frac{d}{dx}(x^2) = 2x$$

—not the same as 1.

Instead, the correct answer is found by the following, the real, correct, *How to Ace Calculus*–certified **product rule:**

$$\boxed{\frac{d}{dx}(fg) = f'g + fg'}$$

It looks a bit messy, but actually it's extremely simple. To say it in words: "The derivative of the product of two functions is the derivative of the first times the second plus the first times the derivative of the second."

Let's apply it to the same case we looked at before, when $f(x) = x$ and $g(x) = x$. Then the rule says,

$$\frac{d}{dx}(x)(x) = (x)'(x) + (x)(x)' = (1)(x) + (x)(1) = 2x$$

That is exactly what the power rule says that the derivative of x^2 should be. So at least in this one case, the rules are consistent. And consistency is what pudding is all about. Mathematics too.

12.4 Quotient rule

Now, we want to be able to take the derivative of a fraction like f/g, where f and g are two functions. This one is a little trickier to remember, but luckily it comes with its own song. The formula is as follows:

$$\frac{d}{dx}\left(\frac{f}{g}\right) = \frac{f'g - fg'}{g^2}$$

How to Remember This Formula (with thanks to Snow White and the Seven Dwarfs) Replacing f by hi and g by ho (hi for high up there in the numerator and ho for low down there in the denominator), and letting D stand in for "the derivative of," the formula becomes

$$D\,\frac{hi}{ho} = \frac{ho\,D(hi) - hi\,D(ho)}{ho^2}$$

In words, that is "ho dee hi minus hi dee ho over ho ho." Now, if Sleepy and Sneezy can remember that, it shouldn't be any problem for you.

As an example,

$$\frac{d}{dx}\left(\frac{x}{x^2+1}\right) = \frac{(x^2+1)(1) - x(2x)}{(x^2+1)^2} = \frac{1 - x^2}{(x^2+1)^2}$$

A Common Mistake Many people remember the quotient rule wrong and get an extra minus sign in the answer. It's very easy to forget whether it's ho dee hi first (yes, it is) or hi dee ho first (no, it's not).

12.5 Derivatives of trig functions

When you were a kid, you had two key facts to remember, your name and your address. If you had confused the two, you might have been lost forever.

You're not a kid anymore, but in this section, you still have just two key facts to remember that you don't want to mix up, namely,

$$\frac{d}{dx}(\sin x) = \cos x$$

$$\frac{d}{dx}(\cos x) = -\sin x$$

The derivatives of all of the other trig functions follow from these.

It is easy to get confused about which of these two derivatives has the negative sign in front. The easiest way to keep it straight is to remember, "Sine keeps its sign when you differentiate." That is to say, when you differentiate the sine function, you do not change the sign for the result. I guess you could remember, "Cosine changes sign," but it's not as catchy.

Showing that the derivatives of the sine and cosine functions are what they are by using the limit definition of the derivative is a little tricky. It uses the fact that we already made a big deal over, namely,

$$\lim_{x \to 0} \frac{\sin x}{x} = 1$$

You should determine whether or not the instructor expects you to be able to derive trigonometric derivatives using this limit.

But anyway, once you know these derivatives, the rest are easy. Say you want to know the derivative of the tangent function. Well,

$$\frac{d}{dx}(\tan x) = \frac{d}{dx}\left(\frac{\sin x}{\cos x}\right)$$

By the quotient rule:

$$\frac{d}{dx}\left(\frac{\sin x}{\cos x}\right) = \frac{(\cos x)(\sin x)' - (\sin x)(\cos x)'}{\cos^2 x}$$

$$= \frac{(\cos x)(\cos x) + (\sin x)(\sin x)}{\cos^2 x}$$

$$= \frac{\cos^2 x + \sin^2 x}{\cos^2 x}$$

Since $\sin^2 x + \cos^2 x = 1$,

$$= \frac{1}{\cos^2 x}$$

and since $1/\cos x = \sec x$,

$$= \sec^2 x$$

This derivative occurs enough that it is probably worth memorizing:

$$\boxed{\frac{d}{dx}\tan x = \sec^2 x}$$

But the derivatives of $\sec x$, $\csc x$, and $\cot x$ are most likely not worth memorizing, since they are easily derived. Again, this depends a lot on the professor. Make sure that you can find the derivatives of these functions using the derivatives of sine and cosine. It makes a typical test question.

And by the way, just as cosine picks up a minus sign upon differentiation, so do the other two trig functions that begin with "c," namely, cosecant and cotangent. So you can just remember, "To avoid a grade of C−, C gets a minus."

12.6 Second derivatives, third derivatives, and so on

This section, in which we differentiate again and again, couldn't be much simpler. We know how to take the derivative of a function already, no problem, easy street. The result, written $f'(x)$, is itself a function of x. But then we can take its derivative again, writing it as $f''(x)$ and calling it the second derivative of $f(x)$. It's like running meat through a grinder a second time before you put the patties on the grill. Makes them mushy and tender, just the way you like them. (For vegetarians, change the analogy to smoothies in the blender.)

For example, if $f(x) = 2x^3$, then $f'(x) = 6x^2$ and $f''(x) = 12x$.

We'll talk about what second derivatives are good for when we discuss graphing.

Now, as Liz Taylor realized after her second marriage, "Why stop at two?" We can keep differentiating over and over.

So in this example, $f'''(x) = 12$ and $f^{(4)}(x) = 0$. Any higher order derivative would also be 0.

In general, $f^{(n)}(x)$ means the nth derivative of $f(x)$.

Example [Trickier]   *What is $f^{(101)}(x)$ if $f(x) = \sin x$?*

You think we're kidding, right? Expecting you to take 101 derivatives? We must be out of our minds.

But you don't really have to take that many derivatives. You see, the derivatives cycle through in foursomes.

$$f(x) = \sin x$$

$$f'(x) = \cos x$$

$$f''(x) = -\sin x$$

$$f'''(x) = -\cos x$$

$$f^{(4)}(x) = \sin x$$

At the fourth derivative, we are back where we started.

Similarly then,

$$f^{(8)}(x) = \sin x$$

$$f^{(12)}(x) = \sin x$$

and in fact

$$f^{(100)}(x) = \sin x$$

since 4 divides 100. So differentiating one more time, $f^{(101)}(x) = \cos x$.

Awe-inspiring, isn't it?

Velocity: Put the Pedal to the Metal

13.1 Velocity as a derivative

Now we need some everyday uses for the derivative. We need to tie it into your day-to-day life, make you feel as if it has some immediacy for you. We want to show how it can become more than just an acquaintance, it can be a real friend. But what makes someone your friend? Well, it helps if he or she does things for you. Sends you cards on your birthday. Makes you a special dinner once in a while.

Derivatives will not do that kind of stuff for you. But they will do things that your other friends can't do, things that will convince you to open your heart. For instance, derivatives will tell you how fast you are going. Say you are driving down the road, and you look at the speedometer, and it reads 65 mph. That speedometer is telling you the instantaneous velocity of the car. It's the instantaneous velocity, because if you lean on the accelerator a little, you will speed up, and the velocity will change. The speedometer is telling you the current speed, not the average speed over the entire trip.

The distinction between constant velocity and varying velocity is important. You know how people always say velocity = distance/time. That's certainly true if you are talking about the average velocity over the entire trip. But unless you have cruise control, you are probably going faster than the average velocity some of the time and slower than the average velocity at other times. Be careful to distinguish between the average velocity, given by

$$\frac{\text{Total distance}}{\text{Total time}}$$

and the instantaneous velocity, which is what we are going to talk about now.

What is the instantaneous velocity? We can think of it as the average velocity over a very short interval of time. So suppose $f(t)$ tells us our position at time t, where we think of our position as a value along a line. (You can think of the line as corresponding to the road going from Walla Walla, Washington, to Pocatello, Idaho.) At a particular time t, we would like to find our instantaneous velocity $v(t)$. Well, let Δt be a very small interval of time. Then $f(t)$ is where we are at time t and $f(t + \Delta t)$ is where we are at time $t + \Delta t$. We have traveled a distance $f(t + \Delta t) - f(t)$ over the time interval Δt. Therefore, our average velocity over the interval of time Δt is

$$\frac{f(t + \Delta t) - f(t)}{\Delta t}$$

Now, we don't want the average velocity over a little time interval, WE WANT TO KNOW THE VELOCITY RIGHT NOW, AT THIS VERY INSTANT. We just take the average velocities as the length of the time interval shrinks to nothing. So

$$v(t) = \lim_{\Delta t \to 0} \frac{f(t + \Delta t) - f(t)}{\Delta t}$$

But hey, you would have to be mighty self-absorbed to not immediately say, "Wait a minute, that's just the limit definition for the derivative. Well, I'll be hornswoggled. That velocity function is just the derivative of $f(t)$."

That's right. $v(t) = f'(t)$, where $f(t)$ is the position function.

13.2 Position and velocity of a car

Example *In driving in a straight line from New York to Boston, your position function given in miles from New York is described by the function*

$$f(t) = \frac{5t^3}{3} - 25t^2 + 120t$$

where t is the number of hours since the trip began. It takes you 8 hours to get to Boston.
(a) *Find your velocity at time t = ½ hour.*
(b) *Do you ever backtrack during the trip?*

Solution (a) First we find your velocity function at any time by differentiating your position function.

$$v(t) = f'(t) = 5t^2 - 50t + 120$$

Then to find your velocity after half an hour, we just plug ½ into the velocity function:

$$v(\tfrac{1}{2}) = f'(\tfrac{1}{2}) = 5(\tfrac{1}{2})^2 - 50(\tfrac{1}{2}) + 120 = \tfrac{5}{4} - 25 + 120 = 96.25 \text{ mph}$$

Hey, it sounds high but a lot of states have upped their speed limits.

Solution (b) Do you backtrack? Yes you do. It seems you left your favorite Mets baseball cap at the rest stop when you took it off to check your thinning hair in the mirror. Passing through Providence, you noticed the shiny spot on your head when you looked in the rearview mirror and realized you had forgotten the cap. Unfortunately, when you got back to the rest stop, you found it in the toilet, where it had been tossed by a Red Sox fan.

How do we see that you backtracked? Well, if the velocity function is ever negative during the trip (that is, when $0 \le t \le 8$), then you must have been backtracking at that point. But notice that

$$v(t) = 5(t^2 - 10t + 24) = 5(t - 6)(t - 4)$$

In particular, $v(t) = 0$ when $t = 4$ hours and $t = 6$ hours. That would be when you realized you lost your hat and turned around, making your velocity go from positive to negative, and also when you turned around at the rest stop after you realized that you could flush the hat goodbye. So we would expect that in between those times, you were backtracking and your velocity was negative. Just to be sure, let's check what the velocity was at time $t = 5$. Then,

$$v(5) = 5(5)^2 - 50(5) + 120 = -5 \text{ mph}$$

Yup, it's negative. You definitely backtracked. You really wanted that hat. But you probably could have made better time if you weren't backing down the shoulder of the northbound lanes. Oh, well...

13.3 Velocity of a falling object

Example   *Ever since you started your calculus class you've suffered from blinding headaches. Nothing helps. Acupuncture, drugs, counseling, you've tried them all, but the headaches get worse and worse. The pain is unbearable, and you decide to end it all. You drive to the middle of the Golden Gate Bridge and climb over the safety rail, 400 feet above the water. With that pain, life is not worth living, so you fling your calculus text (which you carry everywhere) over the edge, and jump out after it. Your height in feet over the water after t seconds is given by the function $h(t) = 400 - 16t^2$.*
(a) *How long till you hit the water?*

Solution (a)   We don't need to do any fancy differentiation for this part. We just need to solve for when the height is zero. Setting $h(t) = 0$ gives

$$400 - 16t^2 = 0$$
$$t^2 = {}^{400}\!/_{16} = 25$$
$$t = 5$$

So you hit the water after 5 seconds.
   But right after you let go of the textbook, your headache disappears. You realize that it is that hated text that has been the cause of all your pain. Suddenly life is a realm of wonder, calling for your presence.

(b) *You had lots of diving lessons when you were a kid. If you are traveling with a velocity of less than 200 feet per second, you can survive the plunge. Will you survive?*

Solution (b)   We need to calculate your velocity when you hit the water. Differentiating $h(t)$ to get the velocity gives

$$v(t) = h'(t) = -32t$$

So the velocity when you hit the water, when $t = 5$, is given by

$$v(5) = -160$$

You hit the water at a speed of 160 feet per second, so you'll make it. The minus sign in the velocity indicates that the height is decreasing. Now, if you can just swim 2 miles in freezing water against a fierce current, you may still be able to make your 3:00 P.M. calculus class.

Chain Rule: S&M Made Easy

With a name like that, you might think that applying the chain rule is about as much fun as hanging from a wrist iron hammered into the wall of a damp dungeon for 20 or 30 years. In fact, it's not so bad—more like 10 years.

The point of the chain rule is to allow us to differentiate composite functions. There are two key ingredients: We need to know how to differentiate functions and we need to know how to compose functions. Hey, since we've already explained those, we're on our way.

There is just a tiny little bit of memorization.

Chain Rule

$$\frac{d}{dx}f(g(x)) = f'(g(x))g'(x)$$

In words, "The derivative of the composition of two functions is the derivative of the outside function evaluated at the inside function times the derivative of the inside function." The thing to pay attention to is the parentheses. Keep your parentheses in the right place and you won't go wrong.

We are not going to prove this result. But don't worry, we wouldn't make up something this crazy.

Look back at that equation again. The information you're supposed to get out of it is how to take the derivative of the composite function $f(g(x))$, the left-hand side of the equation. That means you're supposed to be able to fill in the stuff on the right-hand side of the equation. Let's try it with an example. Take

$$f(x) = \sin x$$

$$g(x) = x^2 + 4$$

Then

$$f(g(x)) = \sin(x^2 + 4)$$

Now, we want to find the derivative of this function. What's $\dfrac{d}{dx}f(g(x))$?

Just fill in the blanks:

$$f'(x) = \cos x$$

$$f'(g(x)) = \cos(g(x))$$

$$= \cos(x^2 + 4)$$

$$g'(x) = 2x$$

So $\dfrac{d}{dx}f(g(x))$ is the product of the last two lines:

$$\frac{d}{dx}f(g(x)) = [\cos(x^2 + 4)]2x$$

The "$2x$" at the end is sometimes called the "tail." As Momma Fox said to Baby Fox when he was leaving for his first day of school, "Honey, don't forget your tail."

There is a second form of the chain rule that is easier to remember but perhaps a little harder to apply. If we want to find the derivative of a composition $f(g(x))$, we first introduce a new variable. We would now like to introduce you to "u." We will let $u = g(x)$. Then we want to find the derivative of $f(u)$ where $u = g(x)$. The chain rule then takes on the easy-to-remember form

$$\frac{df}{dx} = \frac{df}{du}\frac{du}{dx}$$

For example, if we want to find

$$\frac{d}{dx}(\sqrt{x^3 - 7x})$$

we let $f(u) = \sqrt{u}$ and $u = x^3 - 7x$. Then

$$\frac{df}{dx} = \frac{df}{du}\frac{du}{dx}$$

$$= \frac{1}{2\sqrt{u}}(3x^2 - 7)$$

$$= \frac{3x^2 - 7}{2\sqrt{x^3 - 7x}}$$

Of course, neither of these methods is completely trivial to apply. If they were, then why would everyone make such a fuss? So here's the *Real Catch*. Nobody ever tells you what f and g are. They just give you some horribly complicated function and tell you to differentiate it. So the real problem is not composition but *decomposition*—figuring out what the pieces of the function are. Of course, we all learn to decompose eventually, that is, unless we opt for cremation. But why wait until then? Let's do it now.

Here is the most common situation. Suppose that $h(x) = (g(x))^n$. Then $h(x)$ is the composition of $f(x) = x^n$ and $g(x)$, which we know nothing about. But the power rule says $f'(x) = nx^{n-1}$. So

$$h'(x) = f'(g(x))g'(x) = n(g(x))^{n-1}g'(x)$$

This is a generalized form of the power rule, kind of a power rule with a tail. So let's call it the **generalized power rule:**

$$\boxed{\frac{d}{dx}(g(x))^n = n(g(x))^{n-1}g'(x)}$$

Don't confuse this with the general-in-power rule, discussed in our forthcoming book, *How to Seize Control of Small Countries: The Streetwise Guide.* An example of the generalized power rule is

$$\frac{d}{dx}(\sqrt{3x^5}) = \frac{d}{dx}((3x^5)^{1/2}) = \frac{1}{2}(3x^5)^{-1/2}(15x^4)$$

Example (Complicated) *Let $k(x) = (\cos(\sqrt{x} - x))^3$. Compute $k'(x)$.*

This is pretty nasty, but we can see that it is made up of pieces. The $\sqrt{x} - x$ has its cosine taken, and the whole kaboodle is then taken to the third power. So let

$$f(x) = \sqrt{x} - x$$

$$g(x) = \cos x$$

$$h(x) = x^3$$

Then $k(x) = h(g(f(x)))$.

What's $k'(x)$? Regurgitating our knowledge repeatedly, we obtain

$$k'(x) = (h(g(f(x))))'$$

$$= h'(g(f(x)))(g(f(x)))'$$

$$= h'(g(f(x)))g'(f(x))f'(x)$$

We actually had to apply the chain rule to the tail $g(f(x))$ to get from line 2 to line 3 up there. Of course, at this point you might want to ask, "How do I know when I'm done?" The answer is simple: You're done when each $'$ is firmly attached to the letter before it. If you see a $'$ with a parenthesis before it, as occurs, for instance, in the first and second lines above, keep going. Now compute

$$f'(x) = \tfrac{1}{2}(x)^{-1/2} - 1 \qquad ✔$$

$$g'(x) = -\sin x$$

$$h'(x) = 3x^2$$

$$g'(f(x)) = -\sin(\sqrt{x} - x) \qquad ✔$$

$$g(f(x)) = \cos(\sqrt{x} - x)$$

$$h'(g(f(x))) = 3(\cos(\sqrt{x} - x))^2 \qquad ✔$$

Our answer $k'(x)$ is the product of the three lines indicated by checkmarks:

$$k'(x) = [3(\cos(\sqrt{x} - x))^2][-\sin(\sqrt{x} - x)]\left[\tfrac{1}{2}(x)^{-1/2} - 1\right]$$

Before you see if there's still time to drop the class, you should realize that your professor could, theoretically, ask you to take the derivative of a composition of seven functions, each more beautiful than the last. It would require six applications of the chain rule and a whole lot of paper. But it's just not a very practical thing to ask. For that matter, some poor soul would have to correct the papers, and even graduate students are not that fond of tracing through six pages of calculations. Two applications of the chain rule is all a decent human being will ever ask for on an exam. Furthermore, no more than half the class will even be able to do that. So take a look at that last example; you might as well be in the top half.

Graphing: How to Doodle Like an Expert

15.1 Graphing functions

Instead of using the connect-the-dots method, let's see what the derivative tells us about how to graph a function. Suppose we know that the derivative of a function is positive somewhere. [In mathese, that means $f'(x) > 0$ at some fixed x.] So the slope of the tangent line is positive, the graph is positively sloped, and THE FUNCTION IS INCREASING. On the other hand, if the derivative is negative somewhere, it means THE FUNCTION IS DECREASING. (See Figure 15.1.)

So if you go into the boardroom of the Honemacher Tent Company and say, "Ladies and gentlemen, I am sorry to announce that the derivative of our profit function is currently negative," you will see some unhappy campers. Not that this means there aren't currently profits; on the contrary, you're all making lots of money. But it does mean that the profits are decreasing, so someday soon, the profits could sink below zero, and you will all be pumping out septic tanks at the local Jellystone RV Park.

On the other hand, suppose you are an intelligent alien plant spore trying to take over the human race. A lieutenant spore reports, "The derivative of the number of *Homo sapiens* controlled by us is positive." Then you are ecstatic (or at least as ecstatic as a spore can be, which isn't much), because

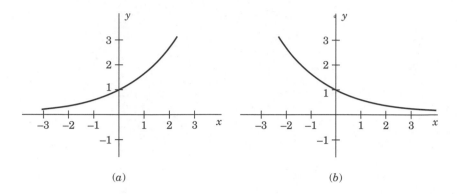

Figure 15.1 (*a*) Function increasing. (*b*) Function decreasing.

it means that the total number of humans that you control is going up, and with any luck you will eventually control everyone on earth. That'll be one more planet down and just a couple of hundred thousand to go before no one ever laughs at plant spores again.

Group Project   Look around the class and estimate how many of your fellow students are controlled by alien plant spores. Professors don't count.

But what if the derivative of a function at a particular value for x is neither negative nor positive, but equal to 0? Then the function is neither decreasing nor increasing. It has plateaued. It may mean we are at a peak (Figure 15.2*a*). It may mean we are at a trough (Figure 15.2*b*). It may just mean we're at a rest stop (Figure 15.2*c*).

But it definitely means that we have a horizontal tangent line at this point. Now THERE is some useful information. Knowing where the plateaus occur is the kind of info that determines the fate of the earth, at least as far as these alien spores go. Let's call the x coordinate of a point where the derivative has value 0 a *critical point*.

Let's State What We Know   If the derivative of a function is defined everywhere, then at a peak or trough of the function,

$$f'(x) = 0$$

We call a peak a *local* (or *relative*) *maximum* of the function and a trough a *local* (or *relative*) *minimum* of the function.

Example   *Find all local maxima and local minima of the function*

$$f(x) = x^2 - 4x + 5$$

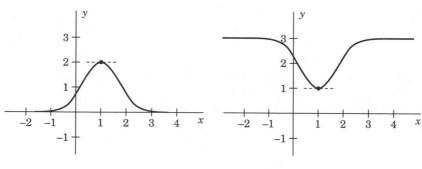

(a) Slope is 0 at a peak.   (b) Slope is 0 at a trough.

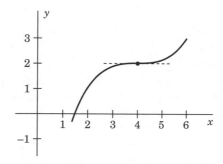

(c) Slope is 0 at a rest stop.

Figure 15.2 Derivative of a function at a particular value for x is equal to 0.

Local maxima and minima can only occur at the values of x where $f'(x) = 0$, namely, at the critical points. But

$$f'(x) = 2x - 4$$

So

$$f'(x) = 0$$

when

$$2x - 4 = 0$$

$$x = 2$$

Notice that for $x < 2, f'(x) < 0$, meaning the function is decreasing. When $x > 2, f'(x) > 0$, meaning that the function is increasing. Since when $x = 2$,

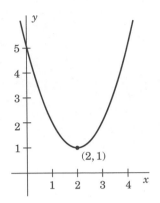

Figure 15.3 Graph of $f(x) = x^2 - 4x + 5$.

$f(x) = 1$, we know that the function passes through the point (2, 1) and that the graph of the function must look like the one shown in Figure 15.3.

Not bad. With very little effort, we have pretty much determined what the graph of the function looks like. In particular, we have seen that it has exactly one local minimum and no local maxima. Pretty good for a few minutes' work. It's accurate and has none of the dronelike boredom of plugging in points and connecting the dots.

Example   *Graph* $f(x) = x^3 - 3x + 1$.

First, we want to find the points where the peaks and troughs can occur, the points where $f'(x) = 0$. Now we start by calculating $f'(x)$:

$$f'(x) = 3x^2 - 3$$

Next we set it equal to 0 and solve:

$$3(x^2 - 1) = 0$$

$$3(x - 1)(x + 1) = 0$$

This gives $x = 1$ or $x = -1$ as the two possible answers.

So potential peaks and troughs occur at $x = 1$ and $x = -1$. Now how do we decide which is a peak and which is a trough, or even if either one is a peak or trough? Maybe they're just rest stops. What's a mother to do? Use the derivative. We look at the sign of the derivative at other values of x to

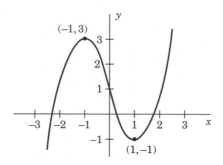

Figure 15.4 Graph of $f(x) = x^3 - 3x + 1$.

determine when $f(x)$ is increasing or decreasing. And how do we decide what the sign of $f'(x)$ is for various values of x? We make a little table of signs for the factors and use them. So for instance, when $x < -1$, $x - 1$ is negative, and $x + 1$ is negative. Since the product of two negatives is positive, we know $f'(x)$ is positive here and $f(x)$ is increasing. Similarly for the other intervals.

Function	$x < -1$	$-1 < x < 1$	$1 < x$
$x + 1$	$-$	$+$	$+$
$x - 1$	$-$	$-$	$+$
$f'(x) = (x - 1)(x + 1)$	$+$	$-$	$+$
$f(x)$	Increasing	Decreasing	Increasing

As we pass through $x = -1$, $f(x)$ goes from increasing to decreasing, so there must be a local maximum at $x = -1$. As we pass through $x = 1$, $f(x)$ goes from decreasing to increasing, so there must be a local minimum at $x = 1$.

To graph the function, we still need to know the y values at $x = -1$ and $x = 1$. They are $f(-1) = 3$ and $f(1) = -1$. The graph must look like Figure 15.4.

15.2 Tricky graphs that can trip you up

Okay, so graphing looks pretty straightforward. What else could come up? Well, we haven't seen examples of all of the problems you might yet encounter. Let's check out a few in passing.

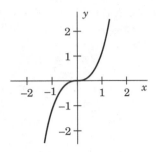

Figure 15.5 Graph of $f(x) = x^3$.

1. A point where the derivative is 0 may not be a local max or min, as in the function

$$f(x) = x^3$$

which looks like Figure 15.5.

2. A function could have a local max or min at a point where the derivative doesn't exist, as happens with

$$f(x) = x^{2/3}$$

Then

$$f'(x) = \tfrac{2}{3}x^{-1/3}$$

At $x = 0$, the derivative isn't defined! However, as we can see in Figure 15.6, $x = 0$ is where the minimum occurs.

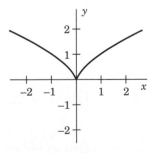

Figure 15.6 Graph of $f(x) = x^{2/3}$. The derivative does not exist at $x = 0$, since the tangent line is vertical.

The points where the derivative is zero OR where it does not exist are the points where the local maxima and minima can occur. All of these x values are called critical points, and it is critical to understand them.

> **Remember for Exam** Local maxima and minima can occur at points where the derivative isn't defined, as well as at points where the derivative equals zero.

15.3 Second derivative test

We've seen that if $f'(x) = 0$ at a point $x = a$, and if $f'(x) > 0$ for $x < a$ (so the function is increasing left of a), and $f'(x) < 0$ for $x > a$ (so the function is decreasing right of a), then $x = a$ must correspond to a local maximum of the function, the peak of a hill. But it's a pain to always check the sign of the derivative on each side of the critical point. Who has the time in today's hectic world? Fortunately there is an alternate way of checking whether we're at a max or min.

We can use the famous **second derivative test.** Here's the amazing fact:

If $x = a$ is a critical point where $f'(a) = 0$, then

1. If $f''(a) > 0$, we have a local minimum at $x = a$.

2. If $f''(a) < 0$, we have a local maximum at $x = a$.

It's that simple. How do you remember which is which? Easy, just use the memory trick shown in Figure 15.7. The happy face has positive sign eyes (for a positive first derivative) and a big old smile that clearly has a local minimum. The sad face has minus sign eyes (for a negative first derivative) and such a sad mouth that it has a local maximum. Of course, these two pictures are incredibly adolescent, and what self-respecting adult would be caught dead relying on them to remember the second derivative test? But

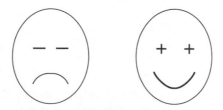

Figure 15.7 At a maximum, the outlook (and the second derivative) is negative. At a minimum, the outlook (and the second derivative) is positive.

remember the bottom line. Swallow your pride or draw them really small in the corner of the page, whatever it takes to keep the rule straight.

Of course, there is one case we left out. What happens when $f''(a) = 0$? Unfortunately, then we don't know squat. We could have a max, min, or a rest stop. We are forced to go back and check for increasing and decreasing on each side of the critical point. Enough talk, let's see some action.

Example *Find the minima and maxima of $f(x) = 2x^3 + 3x^2 - 12x + 14$.*

Hey, this is the same function we saw before. Now we're going to dissect it with our new second derivative kit. We have $f'(x) = 6x^2 + 6x - 12$, and critical points $x = -2$ and $x = 1$. Since $f''(x) = 12x + 6$, we see that $f''(-2) = -18 < 0$, so $x = -2$ is a local maximum. Since $f''(1) = 18 > 0$, $x = 1$ is a local minimum. Talk about the Big Easy.

15.4 Concavity

Now we come to a more subtle aspect of curves. You know those sports car enthusiasts who buy those sleek little roadsters. They don't actually drive the cars. They just keep them in the garage where they go to stroke them. And we can understand why. Running a soft cloth along the fender, caressing the smooth feel of the metal, as its curvature varies, who wouldn't be transported? That indefinable *je ne sais quoi,* the smell of the oil, the coy turn of the bumper ... but enough of this reverie, what's it got to do with math?

Well, math is the ticket to own that rag-top, and once you own it, you will want to measure how fast that fender curves so you can brag about it at the Roadster Rally.

How do we measure those curves? That's a job for the second derivative $f''(x)$. The second derivative is $(f'(x))'$. That is to say, it is the rate of change of the first derivative. In other words, it is the rate of change of the slope of the tangent line. If $f''(x)$ is positive, the slope of the tangent line is increasing, meaning our graph is concave up (Figure 15.8).

Memory Trick Think of "concave up ... pass the cup" to distinguish it from "concave down ... why the frown?" (Okay, so it's trite. Do you want to ace the course or not?)

If $f''(x)$ is negative, that means the tangent line is again changing direction, but this time its slopes are decreasing as we move along the curve, meaning that we have a concave down graph, as shown in Figure 15.9.

The same smiley-face memory devices that worked for the second derivative test work here too.

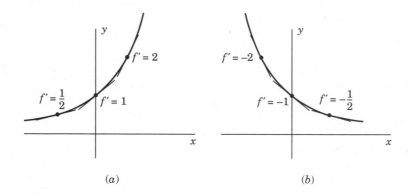

(a)   (b)

Figure 15.8 (*a*) Concave up and increasing. (*b*) Concave up and decreasing.

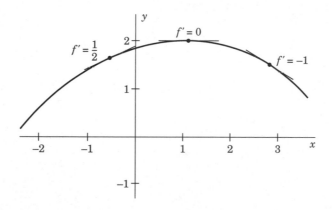

Figure 15.9 $f''(x)$ negative means a concave down graph.

If $f''(x)$ is large and either positive or negative, the slope of the tangent line is changing quickly, and the graph is sharply curved.

If $f''(x)$ is small, the slope of the tangent line is hardly changing at all, meaning that all of the tangent lines are approximately pointed in the same direction. Hence the graph has only the slightest curvature.

At a point where the curve changes concavity from concave up to concave down, we have what is called an *inflection point*. At these points, either $f''(x) = 0$ or $f''(x)$ does not exist. In a typical exam graphing problem, you will be asked to find all inflection points for a function and then draw the graph, labeling concavity. Should be a piece of cake now.

Returning to our example,

$$f(x) = 2x^3 + 3x^2 - 12x + 14$$

since

$$f''(x) = 12x + 6$$

we have a potential inflection point at $x = -\frac{1}{2}$. Calculating shows that

$$f''(x) < 0 \quad \text{for } x < -\frac{1}{2}$$

and

$$f''(x) > 0 \quad \text{for } x > -\frac{1}{2}$$

So $x = -\frac{1}{2}$ IS an inflection point and the graph is concave down to the left of it and concave up to the right.

Maxima and Minima: The Bread and Butter Section

 16.1 **Maxima and minima over closed intervals**

Suppose we have some function that we want to maximize, such as the number of hair follicles on our head. But the number of hair follicles that we ultimately achieve is a function of the amount of dollars x that we are willing to pay Antoine's Hair Transplant Emporium. We have limits on how much we can afford to spend. After all, we have to save some money to have our teeth capped. So we restrict x to $0 \leq x \leq 500$. Now we want to find the absolute (also called global) maximum of the function $H(x)$ giving the number of hair follicles we achieve if we spend x dollars, over the interval $[0, 500]$. Suppose we have an explicit function for $H(x)$ like $H(x) = 12x^2 - 2400x + 120,000$. We can then find the absolute maximum over the interval. First notice that the function and its derivative are defined everywhere over the interval. So at any local maximum (or minimum for that matter), $H'(x) = 0$. Therefore, either the absolute maximum occurs at one of these critical points OR IT COULD OCCUR AT AN ENDPOINT OF THE INTERVAL. In this example, it could occur at 0 or 500. How to decide where the max occurs? Find the critical points, and then evaluate $H(x)$ at each critical point and at each endpoint. Of those values, the one that is the largest is the absolute max.

In this example, $H'(x) = 24x - 2400$. Setting this equal to zero gives us $x = \$100$ as the only critical point. Evaluating $H(x)$ at the critical point and endpoints gives $H(0) = 120{,}000$ hair follicles, $H(100) = 0$, and $H(500) = 1{,}920{,}000$. So the absolute max does occur at the endpoint $x = \$500$. In other words, you don't pay a cent and you have the 120,000 hair follicles you started with, a little on the low side. You shell out \$100 and they begin treatments, enough to make your hair fall out. But if you shell out \$500, wow, what a head of hair you will have, enough to cover your head, neck, *and* back. Think how much you'll save on sunscreen.

16.2 Applied max-min problems

Okay, now we get into the bread and butter section of calculus. Theory is great, and who wouldn't want to spend one's entire life coming up with grand unifying theorems of the universe? But it's an unreliable way to put bread on the table. In this section, we will solve honest-to-god applied questions, like "How can I make a box with the largest possible volume from a single rectangular piece of cardboard?" You may scoff, but this is one of the classic problems, occurring in more calculus books than almost any other single problem. And who can forget the beloved "If a farmer has 100 feet of fence and wants to make a rectangular pigpen, one side of which is along an existing wall, what dimensions should be used in order to maximize the area of the pigpen?"

We know, you're thinking that the number of Americans who make their living from traditional farms is now such an infinitesimally small percentage of the population, it is invisible to the naked eye. So this problem is old-fashioned and irrelevant, harking back to an era when your newspaper was delivered by a kid on a bike with a playing card slapping his spokes. Nowadays, pigs are "produced" and "harvested" by megaconglomerates who own the entire states of Nebraska, Kansas, and Iowa. And newspapers are delivered by guys wearing too tight tee-shirts driving big rusty Oldsmobiles with six-packs hidden under the dash.

But remember that those same agribusiness corporations will pay top dollar to the consultant who can tell them how best to construct their pigpens. And the newspaper delivery conglomerates will be very interested in minimizing the route that the Oldsmobile has to take in order to deliver the papers.

So what theory do we have to work with?

Remember, we said that when a function has a local maximum or minimum, the derivative is 0, if it exists. So the *global* maximum or minimum of a function occurs either at a point where the derivative is 0 or where the derivative doesn't exist (these two kinds of points are critical points) or at an endpoint of the domain.

Warning The maximum or minimum can occur at an endpoint. For example, the silly problem, "Given 100 feet of fence, what are the dimensions of a rectangular pen with *minimal* area?" has solutions at the endpoints of the interval [0, 100]!

Let's start with a simple but common type of example, explaining the general technique for solving max-min problems as we go along.

Problem   *Find two nonnegative real numbers that add up to 66 and such that their product is as large as possible.*

(It is possible to solve this problem without using calculus. But hey, this is a calc course, and we're doing this problem to learn a technique of calculus that can be used on lots of other problems, not to attempt to prove that we can scrape through some kind of miserable existence without needing calculus. So, definitely use calculus.)

Step 1. Name the variables. Let x be the first number, and y the second. Since they are both nonnegative real numbers,

$$x \geq 0 \quad y \geq 0$$

Notice that x and y have to be less than or equal to 66.

Step 2. Write down the function that is to be maximized or minimized, giving it a name. We will use the letter P to denote the function we want to maximize. (Hey, it's a good choice, P stands for product.) Let

$$P = xy$$

We want to maximize this product.

Step 3. Write down any relations between the variables. The variables are related by

$$x + y = 66$$

Step 4. Reduce the function we are maximizing or minimizing to one variable using the relation in step 3. Since $y = 66 - x$, we have that

$$P = x(66 - x)$$

Step 5. Find the critical values of P.

$$P = 66x - x^2$$

so

$$P' = 66 - 2x$$

$P' = 0$ when

$$66 - 2x = 0$$

$$x = 33$$

So we have a critical point. Does P have a maximum or a minimum at this value of x? Maybe neither? Well, it's time to move on to step 6.

Step 6. Apply the second derivative test.

$$P'' = -2$$

So

$$P''(33) = -2 < 0$$

So this is a maximum. A quick check at the endpoints $x = 0$ and $x = 66$ shows there is no competition for King of the Hill, and it gives the maximum for the whole function. Therefore $x = 33$ and $y = 66 - 33 = 33$.

Step 7. Write down the final answer. You would be amazed how often students forget this one.

Our two numbers are 33 and 33.

Now I know what you are thinking. You are thinking, hey, I could have guessed that answer. Good for you. You may be able to get a job answering phones at the Psychic Hotline. Notice also that if the question had asked for two nonnegative numbers that added up to 66 and had the *smallest* possible product, the answer would have been at an endpoint of the domain, namely, 0 and 66.

On to our next problem.

Problem *You are in a rowboat on Lake Erie, 2 miles from a straight shoreline, taking your potential in-laws for a boat ride. Six miles down the shoreline from the nearest point on shore is an outhouse. You suddenly feel the need for its use. It is September, so the water is too cold to go in, and besides, your in-laws are already pretty unimpressed with your "yacht." It wouldn't help matters to jump over the side and relieve your distended bladder. Also, the shoreline is populated with lots of houses, all owned by people who know your* *parents, and would love to get you in trouble with them. If you can row at 2 mph and run at 6 mph (you can run faster when you don't have to keep your knees together), for what point along the shoreline should you aim in order to minimize the amount of time it will take you to get to the outhouse?*

See, we told you calculus would be useful. And here is a problem that could not be of a more immediate or intimate nature. YOU REALLY CARE ABOUT THE SOLUTION.

Step 1. Draw a picture and name the variables. In the last problem we didn't need to draw a picture, since there was no relevant picture. But here, there is. Don't make it too graphic. Leave out the details of the in-laws. Draw a (straight) shoreline, a boat 2 miles offshore, and an outhouse 6 miles down the shoreline. Let x be the distance along the shore from the closest point to you to the point where you land. Then $6 - x$ is the remaining distance you will have to run along the shore to get to the outhouse. See Figure 16.1.

Step 2. Figure out the function that is to be minimized. By the pythagorean formula, the distance you'll have to row is

$$D_w = \sqrt{4 + x^2}$$

where w stands for water. To find the time you'll be in the boat rowing, we can use the fact that

$$\text{distance} = \text{speed} \times \text{time}$$

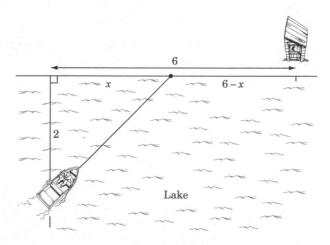

Figure 16.1 How to get to shore, that is the question.

Solving for time,

$$\text{time} = \frac{\text{distance}}{\text{speed}}$$

So the time that you are in the boat on the water will be

$$T_w = \frac{\sqrt{4 + x^2}}{2}$$

On the other hand, the distance you have to run is $6 - x$ miles, at a speed of 6 mph. That means that the time you'll be running is

$$T_L = \frac{6 - x}{6}$$

where L stands for land. The total time for your trip to the outhouse is

$$T(x) = T_w + T_L = \frac{\sqrt{4 + x^2}}{2} + \frac{6 - x}{6}$$

Notice that if you row straight to the nearest point on shore and then hoof the full 6 miles, it will take you $T(0) = 2$ hours. On the other hand, if you decide to row straight to the outhouse, then $x = 6$ and it will take you $T(6) \approx 3.16$ hours. So you'd better hope you can do a lot better than either of those.

Step 3. Write down any relations between the variables. In this example, we have only one variable present, so we can take this step off and relax for a couple of seconds. Okay, off we go.

Step 4. Reduce to one variable. Already done. More relaxing. Now, let's get to it.

Step 5. Find the critical points. Take the derivative,

$$T(x) = \frac{(4 + x^2)^{1/2}}{2} + \frac{6 - x}{6}$$

$$T'(x) = \left(\frac{1}{2}\right)\frac{(4 + x^2)^{-1/2}}{2} 2x - \frac{1}{6}$$

and set it equal to 0:

$$\frac{(4 + x^2)^{-1/2} 2x}{4} - \frac{1}{6} = 0$$

Then solve:

$$\frac{x}{2(4 + x^2)^{1/2}} = \frac{1}{6}$$

$$6x = 2(4 + x^2)^{1/2}$$

$$3x = (4 + x^2)^{1/2}$$

$$9x^2 = 4 + x^2$$

$$8x^2 = 4$$

$$x^2 = \frac{1}{2}$$

$$x = \frac{1}{\sqrt{2}} \approx 0.707 \text{ mile}$$

Steps 6 and 7. Check to make sure it's a minimum.

Notice that we get only one critical point between $x = 0$ and $x = 6$. So we expect this to be the minimum. We could apply the second derivative test to make sure, but it really isn't necessary. As we will see, our total time will be less than for either of the two extremes $x = 0$ or $x = 6$, so this must be a minimum. For $x = 1/\sqrt{2}$, it will take you a total time of:

$$T\left(\frac{1}{\sqrt{2}}\right) = \frac{\sqrt{4 + (1/\sqrt{2})^2}}{2} + \frac{6 - 1/\sqrt{2}}{6}$$

If you calculate this number out, you find that

$$T\left(\frac{1}{\sqrt{2}}\right) \approx 1.94 \text{ hours}$$

which is less than either 2 or 3.16 hours, but it's still quite a while to hold out. Well, good luck. Maybe next time, you'll plan a head.

Problem   *Northern Kansas University is building a new running track. It is to be the perimeter of a region obtained by putting two semicircles on the ends of a rectangle as in Figure 16.2. However, due to financial constraints, the administration has decided to grow corn in the area surrounded by the track. If the track is to be 440 yards long, determine the necessary dimensions to build the track in order to maximize the area for growing corn.*

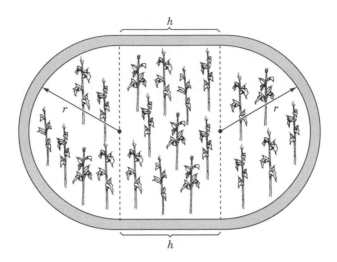

Figure 16.2 Running around the cornfield.

Step 1. Draw the picture. We did this for you. See, we really are on your side.

Step 2. Figure out the function to be maximized. The problem asks you to maximize the area, so you need to write down what the area is in terms of r and h. The area A is the sum of the areas of the semicircular regions (each of area $\pi r^2/2$) and the rectangular region (of area $2rh$):

$$A = \pi r^2 + 2rh$$

Step 3. Write down any relations between the variables. Here you have to use more information from the problem. The 440 has to be good for something. In this case, it gives you a relation between r and h since the perimeter of the track will be the circumference of the two half-circles, which contributes a total of $2\pi r$, and the length along the straightaways, which is $2h$:

$$440 = 2\pi r + 2h$$

Step 4. Reduce the function to be maximized to one variable. Solve the equation $440 = 2\pi r + 2h$ for h:

$$h = 220 - \pi r$$

and use this to write A as a function of one variable:

$$A = \pi r^2 + 2r(220 - \pi r)$$

so

$$A(r) = 440r - \pi r^2$$

Step 5. Take the derivative and set it equal to 0.

$$A'(r) = 440 - 2\pi r$$

Set it equal to zero:

$$440 - 2\pi r = 0$$

$$r = \frac{220}{\pi}$$

Step 6. Use the second derivative test.

$$A''(r) = -2\pi$$

so A'' is negative no matter what value of r you stick in. So we must have just found a maximum. You can check the values of A at the endpoints if you're nervous.

Step 7. Answer the question. This problem is a good example of why this last step is so important, because until you do this, you're missing what your professor thinks is the punchline. The question was: *What are the dimensions of the field?* So far we've only found *one* of the dimensions, r. What's h? It's easy to find, since

$$h = 220 - \pi r$$

so

$$h = 220 - \pi \left(\frac{220}{\pi} \right) \qquad h = 0$$

Aha! you're supposed to exclaim, the track is circular! How amazing! (Even if you personally remain deeply unenthusiastic about the ultimate shape of the track, remember to finish the problem anyway. Circle those r and h values, in purple if possible.)

Now that we are getting fluent at this, we won't delineate each step individually. Let's try that old chestnut, the pigpen problem.

Problem (Pigpen Problem)   *If a farmer has 100 feet of fence and wants to make a rectangular pigpen, one side of which is along an existing straight fence, what dimensions should be used in order to maximize the area of the pen?*

Figure 16.3 Pigpen.

Now, don't get distracted by the pigs. They are just getting in the way. Concentrate on the pen. Let's draw it. That's the catchphrase for these problems. DRAW, DRAW, DRAW. What do we get? See Figure 16.3.

The total area is $A(x) = x\left(\dfrac{100 - x}{2}\right) = \dfrac{100x - x^2}{2}$. The values for x have to be in the interval [0, 100], since we only have 100 feet of fence to work with. Then $A'(x) = \dfrac{100 - 2x}{2} = 50 - x$.

Setting $A'(x) = 0$, we have $50 - x = 0$, so $x = 50$. At the endpoints, the area is 0, but at $x = 50$, the area is 1250 square feet. So $x = 50$ must be the correct value for x. Also notice that $A''(x) = -1$, meaning that the second derivative verifies that this is a local maximum. Of course we are not done with the problem until we say: "The dimensions that maximize the area of the pen are 50×25 feet."

What other common types of problems are there? There is the notorious box problem: Make an open-topped box by cutting four square corners out of a rectangular sheet of cardboard (of fixed dimensions, say 5×8 feet) and then folding up the four sides. Find the dimensions of the squares to be cut out in order to maximize the volume of the box. It's a classic. Try one like this at home.

MAXIMIZING PROFIT PROBLEMS

The last common type of problem is the business problem. Now, let's be honest here. A lot of what drives us all is the desire for money. Sure, there's sex, hunger, romance, and TV, but money is the big chimichanga. So here we are halfway through the book, and we have yet to tell you how to make mountains of money. Where do we explain how to become so incredibly wealthy that you don't give away money because your arm would get tired? Well, this is it. The section is finally here. So let's get rich.

Problem (Exploiting Humanity Problem)   *You have just invented a new peanut butter guacamole dip, and you open a stand in front of the student union to sell this goop by the jar. Somehow a rumor gets started, certainly not traceable back to you, that it is an aphrodisiac, and sales take off. At a price of $1.00 a jar, you sell 500 jars a day. For each nickel that you increase the price, you sell two fewer jars. Assuming that your fixed cost per day is $200 (protection money), and the cost per jar to you is 50 cents, determine the price for which you should sell your dip in order to maximize your profit.*

Since you know how to finish off a problem like this, let's just set it up. Let $P(x)$ be your daily profit, where x is the price in dollars per jar. Let y be the number of jars that you sell per day. Then y depends on the price x that you set. Each $1 increase in x above your initial price decreases your sales by 40 units. So the daily sales number y at price level x is given by the equation

$$y = 500 - 40(x - 1)$$

$$= 540 - 40x$$

Your daily profit will be given by taking revenue − cost. Revenue is given by xy and cost by $200 + 0.5y$. So profit is given by

$$P(x) = xy - 200 - 0.5y = (x - 0.5)y - 200$$

Substituting in our function for y in terms of x we find

$$P(x) = (x - 0.5)(540 - 40x) - 200$$

You can take it from here, finding $P'(x)$, setting it equal to zero, and locating first the critical points and then the optimal price. You're on your way to a lucrative career.

Implicit Differentiation: Let's Be Oblique

Suppose your boss says, "I have had it with your incompetence. You've screwed up everything we've ever given you to do. The entire company is on the brink of bankruptcy, single-handedly thanks to you. Now, clean out your desk and don't show your face here again!"

You might argue that the boss never actually said, "You're fired." There was no explicit dismissal ever stated. But you have to face the facts: It was certainly implied.

Now an equation can also have an implicit meaning. For instance, the equation

$$y + x^2 - 3 = 0$$

implicitly describes y as a function of x. The explicit description of y as a function of x would come from moving $x^2 - 3$ to the other side of the equation to obtain $y = -x^2 + 3$.

More generally, an equation can implicitly give y as a function of x, even if we cannot write y explicitly as a function of x by isolating y on one side of the equation. For instance,

$$y^5 + y + x^7 + 2x = 0$$

can be thought of as giving y as a function of x, even though we can't write down the explicit formula for y. Given a value of x, we plug it into the equation and then solve to find the corresponding value of y. That's how y depends on x.

Since we can implicitly define y as a function of x by an equation, we would like to be able to differentiate this new function of x. We want to be able to find its rate of change. But since we don't know the function explicitly, we will just have to differentiate the entire equation with respect to x. As we do so, we will treat x as usual, but will think of y as a function of x. In particular, its derivative will be written $\dfrac{dy}{dx}$.

Example   *Find $\dfrac{dy}{dx}$ if y is defined implicitly as a function of x by*

$$y^2 + xy + 3x = 9$$

We solve this by differentiating with respect to x, while treating y as a function of x. We obtain

$$2y\frac{dy}{dx} + 1 \cdot y + x \cdot \frac{dy}{dx} + 3 = 0$$

Pay attention to what happened when we differentiated y^2. We obtained $2y\dfrac{dy}{dx}$ by using the chain rule. And when we differentiated xy, we obtained $1 \cdot y + x \cdot \dfrac{dy}{dx}$, by applying the product rule.

Now, we just solve this equation for $\dfrac{dy}{dx}$:

$$2y\frac{dy}{dx} + x\frac{dy}{dx} = -y - 3$$

$$(2y + x)\frac{dy}{dx} = -y - 3$$

$$\frac{dy}{dx} = \frac{-y - 3}{2y + x}$$

And there it is, $\dfrac{dy}{dx}$.

Now there's a final bit of work to do to actually get a number giving the slope of the curve at a given point. We need to plug in not just the x value, but also the y value. So to find the derivative $\dfrac{dy}{dx}$ at the point (2, 1) on the curve, we plug in to get

$$\frac{dy}{dx}(2,\ 1) = \frac{-1-3}{2\cdot 1+2} = \frac{-4}{4} = -1$$

Related Rates: You Change, I Change

Oh, yes, these problems can be nasty. Lots of students fear related rates problems. Why? Maybe because they are word problems, and students just don't like word problems. Having to change English into mathematics intimidates many people. It's as if when they hear it in words, the mathematical sides of their brains shut down.

But related rates problems aren't really so bad. There are just a few species in this phylum. In this chapter, we will dissect the three major types.

So just what is a related rates problem? It is a problem that has an equation relating two or more things which change over time, and we want to find the derivative of one of the functions at a particular time.

Here's a boring but simple example.

Example *Suppose that both x and y depend on t (thought of as time) and for all values of t, $\sin x + \cos y = 1$. Suppose the most important thing in our lives right now is to figure out what dy/dt equals when $x = \pi/6$, $y = \pi/3$, and $dx/dt = 2$.*

In these problems we always have information about the other quantities (x, y, and dx/dt in this case), called the particular information.

To solve, we just differentiate the entire equation involving x and y with respect to t using implicit differentiation. Keep in mind here that, since we are differentiating with respect to t, both x and y must be treated as functions of t:

$$\sin x + \cos y = 1$$

$$\cos x \frac{dx}{dt} - \sin y \frac{dy}{dt} = 0$$

Now we plug in the particular info that we have:

$$\left(\cos \frac{\pi}{6}\right) 2 - \left(\sin \frac{\pi}{3}\right) \frac{dy}{dt} = 0$$

$$\frac{\sqrt{3}}{2}(2) - \frac{\sqrt{3}}{2} \frac{dy}{dt} = 0$$

Next we solve for $\frac{dy}{dt}$:

$$\frac{dy}{dt} = 2$$

Pretty easy. Of course, that wasn't really a word problem, so we didn't have the extra job of figuring out how to write equations for all the quantities described. But these problems are all this easy once you have translated them into the right mathematical equations.

We now give a recipe for solving related rates word problems. It turns out that this recipe is so powerful it can also be used to make chicken soup. Here it is.

A Recipe for Solving Related Rates Problems, and for Making Chicken Soup

Step 1. Read the problem. Hey, we're not kidding. You would be amazed how many people skip this step.

Step 1 for soup: Read the instructions on the can. You'd be amazed how many people skip this step too. The soup company paid a team of gourmet cooks, college graduates, and marketing researchers millions of dollars to write three lines that take 10 seconds to read, and almost nobody reads them.

Step 2. Get a pen and draw a diagram of what is going on. Always draw the diagram in the general case, without the particular information. Give names to all the relevant quantities.

Step 2 for soup: Get a pan and draw the soup from the can into the pan. If you spill it all on the floor, clean up the mess and start over.

Step 3. Find the equation that gives the general relationship between the variables you just named. Use your diagram to help you do this.

Step 3 for soup: Find the knobs that turn on the burner of the stove that's under the pan. Usually there is a diagram on the stove that shows what knob to turn.

Step 4. Find the particular information corresponding to the particular time that you are interested in. Write it down in a box labeled "Particular Info." Also in this box, put down the quantity you are after (the derivative of one of the variables at this time), followed by a question mark.

Step 4 for soup: Turn the particular knob that turns on the particular burner on which you have put the soup pot.

Step 5. Differentiate the general equation implicitly with respect to time. The resulting equation will contain at least two derivatives.

Step 5 for soup: Heat the soup for the indicated amount of time.

Step 6. Plug the particular information into the resultant equation (this step must come AFTER step 5!) and then solve for the unknown quantity you were after.

Step 6 for soup: Plug the stove into the electric outlet if the burner hasn't gotten hot after a few minutes. On the other hand, if you smell gas, forget about plugging in the stove. It's not electric. Run for the door.

Step 7. Write down your answer and circle it in purple. Then have some chicken soup. It can't hurt.

Step 7 for soup: Eat the soup already! What, you don't like it maybe?

Okay, let's try one.

First, we'll look at perhaps the most prevalent problems of this type. We call this category **similar triangles related rates problems.**

Example (Bigfoot Problem)   *Bigfoot wanders out of the mountains and onto a city street in Seattle. Weighing in at 441 fur-covered pounds and reaching a height of 8 feet, he looks like an unusually big grunge rocker. Suppose that the Big Guy, new to the Big City, walks quizzically toward a streetlight at a rate of 2 feet per second. If the light is at a height of 12 feet, at what rate is the length of Bigfoot's shadow changing when he is 6 feet from the base of the lamppost?*

All right, before you get all upset, we admit it's not too clear why we would want to solve this problem. Exactly who is it that is supposed to care about the rate at which Bigfoot's shadow is growing? Given that we can infer that it's dark and there is a big hairy creature about to destroy the only street light around, after which we will all be plunged into a darkness that this beast considers the perfect condition for hunting, why are we standing on the corner with pad and pencil, worrying about his shadow? Good question.

But this is mathematics. And in mathematics, the fake examples are sometimes more educational than the real ones. Besides, once we get good at a problem like this, we can solve it in a couple of seconds, and then get the hell out of there, before he ever reaches the streetlight.

So let's do it in steps.

Step 1. Read the problem. Maybe read it more than once. Notice that some of the information is completely irrelevant, like Bigfoot's weight.

Step 2. Make a diagram such as that shown in Figure 18.1. This doesn't require a detailed picture of the Sasquatch, but if you have the time, feel free. Remember that the diagram is supposed to represent the general situation. At all times, Bigfoot has a height of 8 feet and the light is at a height of 12 feet, so we can mark those heights on the diagram. What varies in the general case? Well, Bigfoot is walking toward the lamppost, so the distance from him to the base of the post is changing. We'll label that distance x. Let's not worry about which part of him we measure from.

Of course, we are interested in the shadow, so let's label the length of the shadow y. What about the edges that we haven't labeled with anything? Should we make up variables and label those edges?

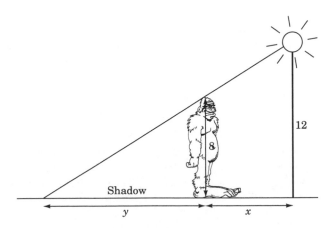

Figure 18.1 Bigfoot discovers electricity.

Actually, no. Those edges don't play a role in the calculations, and having extra variables will just confuse the issue. And remember, we have a tight schedule here. We don't have a lot of extra time to blow on this problem.

Step 3. Find the general equation relating the variables. Notice that there are two right triangles in the picture, one with base y (the shadow) and height 8 (the Big Fellow), and the other with base $x + y$ (from the tip of the shadow to the base of the lamppost) and height 12 (the lamppost). These two triangles are similar triangles, as their angles are the same. Therefore, the ratios of their corresponding edges should be equal:

$$\frac{8}{y} = \frac{12}{x + y}$$

$$8(x + y) = 12y$$

$$8x = 4y$$

$$2x = y$$

Step 4. Write down the particular information. So, we want to know the rate of change of the length of the shadow when the Abominable Snow Guy is 6 feet from the lamppost, that is, when $x = 6$. The rate of change of the length of the shadow is the derivative of y. So we want to know $\frac{dy}{dt}$ when $x = 6$. Since he is moving toward the light at 2 ft/sec, we know $\frac{dx}{dt} = -2$. Don't forget to label the box with "Particular Info."

$$\boxed{\text{Particular Info:} \quad x = 6, \quad \frac{dx}{dt} = -2, \quad \frac{dy}{dt} = ?}$$

Step 5. Differentiate the general equation implicitly.

$$2\frac{dx}{dt} = \frac{dy}{dt}$$

This is an unusually simple implicit differentiation.

Step 6. Plug in the particular info.

$$2 \cdot (-2) = \frac{dy}{dt}$$

So $\dfrac{dy}{dt} = -4$. When Bigfoot is 6 feet from the base of the light, his shadow is shortening at a rate of 4 feet per second.

And that's the answer!

Step 7. Circle it in purple.

Let's take a look at an example of another type of related rates problem, the type of problem where a substance is being put into or taken out of a container, and we want to relate the rate of change of the volume to the rate of change of some other quantity like the height or radius. We will call problems of this type **wine barrel problems.**

Example (Wine Barrel Problem)   *Suppose the wood nymphs and satyrs are having a hot party in honor of Bacchus and the wine is flowing freely from the bottom of a giant cone-shaped barrel which is 12 feet deep and 6 feet in radius at the top. If the wine is disappearing at a rate of 6 cubic feet per hour, at what rate is the depth of wine in the tank going down when the depth is 4 feet?*

Step 1. Read the problem. Let's assume you just did that.

Step 2. Draw a picture of the cone-shaped barrel (Figure 18.2). The wood nymphs and satyrs are optional. The height and radius of the entire barrel can be labeled, since they don't vary in the problem. As the

Figure 18.2 Wood nymphs and satyrs frolic.

party progresses, the depth of the wine in the barrel and the radius of the surface of the wine *will* vary. So we'll label these with an h (for height) and an r. That's everything we need to label in the picture.

Step 3. Find the general equation relating variables. This is a little tricky. Ultimately we want to relate the rate of change of the volume to the rate of change of the depth. So we want an equation that relates volume to depth. However, volume doesn't even show up in our diagram.

But we do know that the volume of a cone V is given by $V = \pi r^2 h /3$ (which is easy to remember as one-third the volume of a cylinder with height h and radius r).

Unfortunately for us, this gives the volume in terms of the two variables r and h rather than just in terms of h. But our diagram does show a nice relationship between r and h, and we can use this to get rid of r in our equation for the volume.

Notice that we again have similar triangles, and therefore we see $r/h = 6/12$. Hence, $r = h/2$. We plug that into the general equation for the volume to obtain

$$V = \pi \left(\frac{h}{2}\right)^2 \frac{h}{3},$$

so

$$V = \frac{\pi h^3}{12}$$

This is our general equation.

Step 4. Write down the particular information at the time we are interested in. We want to know $\dfrac{dh}{dt}$ when $h = 4$. The volume is going down at a rate of 6 cubic feet per hour throughout the problem, so $\dfrac{dV}{dt} = -6$ (negative since the volume is decreasing).

$$\boxed{\text{Particular Info:} \quad h = 4, \quad \frac{dV}{dt} = -6, \quad \frac{dh}{dt} = ?}$$

Step 5. We implicitly differentiate the general equation to get

$$\frac{dV}{dt} = \pi \frac{3h^2}{12} \frac{dh}{dt}$$

(We are using the chain rule here.)

$$\frac{dV}{dt} = \pi \frac{h^2}{4} \frac{dh}{dt}$$

Step 6. Plug in particular info and solve.

$$-6 = \pi \frac{4^2}{4} \frac{dh}{dt}$$

$$\frac{dh}{dt} = \frac{-3}{2\pi} \approx -0.477 \text{ foot per hour}$$

So at the particular time that we are interested in, the wine level is dropping by just a hair less than half a foot per hour. Somebody's drinking more than his or her fair share...

The third type of related rates problems is the **Pythagorean theorem type problem.**

Example (Tracking a UFO)   *The TV show "Unnatural Experiences and Nauseating Aliens" claims to have photographed a UFO as it flies overhead, at a constant altitude of 3 miles and a constant speed. They state that they took the photo (which resembles a very blurry cigar, including what appears to be the cigar label) at the instant after the UFO had already flown horizontally 4 miles past the point directly above them. They say that the radar gun showed that at the instant the photograph was taken, the distance between the photographer and the UFO was growing at a speed of 384 mph. Determine the speed that the UFO/cigar was traveling.*

Step 1. That's right. Read the problem.

Step 2. Draw the picture (Figure 18.3). We will label the horizontal distance x and the actual distance h (for hypotenuse). Avoid the tremendous temptation to label x with the distance at the time the photo was taken.

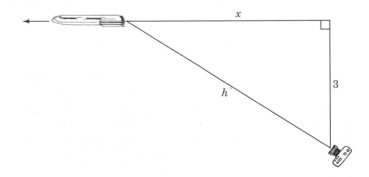

Figure 18.3 UFO captured on film.

Remember to make this diagram represent the general situation at no specific time.

Step 3. Find the general equation. Thank goodness for good old Pythagoras, who, for a guy dressed in a sheet, had surprising insight about right triangles. Since we have a right triangle, we know that $3^2 + x^2 = h^2$.

Step 4. We want to determine the particular info at the time we are interested in, namely, when the photo was taken. At that time, $x = 4$. Using the general equation, we can also determine h at this time:

$$3^2 + 4^2 = h^2$$

So $25 = h^2$ and $h = 5$. We also know the rate of change of h at the particular time, namely, $\dfrac{dh}{dt} = 384$ mph. So, we have

> Particular Info: $x = 4,\quad \overset{h}{\cancel{y}} = 5,\quad \dfrac{dh}{dt} = 384,\quad \dfrac{dx}{dt} = ?$

Step 5. We implicitly differentiate $9 + x^2 = h^2$ to get

$$2x\frac{dx}{dt} = 2h\frac{dh}{dt}$$

or

$$x\frac{dx}{dt} = h\frac{dh}{dt}$$

Step 6. We plug in the particular info to get

$$4\frac{dx}{dt} = 5(384)$$

$$\frac{dx}{dt} = \frac{5(384)}{4} = 480 \text{ mph}$$

Wow! That is one fast stogie!

Addendum: When scientists pointed out that a UFO was unlikely to have the words El Corona written on its hull, and they asked to speak to the photographer, the producers of the show explained that the aliens had resented the intrusion on their privacy and had turned the photographer into a puddle of paparazzi sauce. And so ends another close encounter of the third kind.

The vast majority of related rates problems will fall into one of these three categories.

Differential: Estimating Your Way to Fame and Glory

Being able to make quick estimates is essential for making your way in the world. When the the gas gauge hits E, you want to be able to estimate the number of additional miles you can drive before the car coughs to a halt. When the elevator doors open, you want to be able to tell from a quick glance at the crowd if you will put the total weight over the 2000-pound limit. And you want to be able to make a quick estimate of compatibility, in case your new friend from calculus class asks you out for the first time, and it's for a two-week trip to Majorca.

Similarly, you want to be able to estimate the values of functions at places where you don't know the exact value. Let's face it, this is almost everywhere. Lots of functions are hard to compute at most places. For instance, try to think off the top of your head what $\sqrt{4.1}$ is, or $\sin(44°)$. Hey, don't feel bad. We don't know what they are, either.

Of course, for such familiar functions we could always punch a few keys on a calculator, and it would tell us the values out to eight decimal places. That leaves you wondering how the calculator knew the answer. You might think the calculator has the values of all of the basic functions memorized. But just imagine how many different values for x you can put in your calculator. Billions and billions. The calculator does not have billions and billions of values of $\sqrt{x}$ and $\sin x$ memorized. Calculator memory is too limited for that.

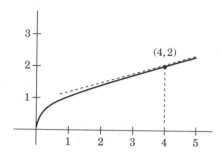

Figure 19.1 Square root function.

Instead, the calculator has a system for estimating each answer. You punch in your request, the calculator does a computation that yields an estimate for the answer, and then it makes sure the estimate is accurate to eight decimal places.

So we will look at a method for estimating a value like $\sqrt{4.1}$. Notice that we do know the value of $\sqrt{x}$ for an x near 4.1, namely, $\sqrt{4} = 2$. We'll take this as our first estimate and then add a slight correction, to get our answer closer to the actual value of $\sqrt{4.1}$. Let's look at the graph shown in Figure 19.1.

The value of the function $f(x) = \sqrt{x}$ is 2 when $x = 4$ and something slightly larger than 2 when $x = 4.1$. How much larger than 2? Well, we could take the tangent line at $x = 4$ and extend it out to $x = 4.1$. The tangent line is pretty close to the original graph of $f(x) = \sqrt{x}$ near $x = 4$, and so we could just add the additional height coming from the tangent line to the value of $f(x)$ at 4 to get a damn good approximation to the actual value of the function. So how much additional height do we pick up? We have a line with some slope, and we know by how much x changes, namely, 0.1. Therefore we can find out by how much y changes, since

$$\frac{\text{Change in } y}{\text{Change in } x} = slope$$

And what is the slope of this line? Hey, it's the derivative of $f(x) = \sqrt{x}$ at 4. Since

$$f'(x) = \frac{1}{2x^{1/2}}$$

$$f'(4) = \frac{1}{4}$$

So if we travel along the tangent line from $x = 4$ to $x = 4.1$, our height will increase by a y value of $(0.1)(\frac{1}{4}) = 0.025$. That means we expect that

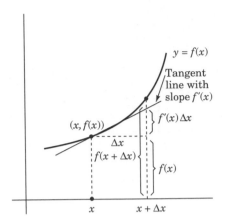

Figure 19.2 Approximating $f(x + \Delta x)$.

the value of $\sqrt{4.1}$ is about $2 + 0.025 = 2.025$. If you check with a calculator, $\sqrt{4.1}$ turns out to be 2.0248 rounded to four decimal places. In other words, our approximation is incredibly, amazingly close to the actual value. In some countries, this counts as a miracle.

Let's look at the general problem of approximation now. Suppose that we have a function $y = f(x)$ and we know its value at a particular point x, but we are really interested in its value at a slightly different point $x + \Delta x$. Think of Δx as a very small quantity, just like the 0.1 in the previous example. So we want to find an approximation for the value of the function f at $x + \Delta x$. That is, we want to find an approximation for $f(x + \Delta x)$.

At x, we know that the value of the function is $f(x)$, a quantity that we are assuming we can compute. Again, we will use the tangent line to the graph of $y = f(x)$ at the point $(x, f(x))$ in order to obtain our approximation (see Figure 19.2). The slope of the tangent line is $f'(x)$, so the amount of height we want to add to $f(x)$ should be that slope times the change in x, which is $f'(x)\,\Delta x$.

So we have the result

$$f(x + \Delta x) \approx f(x) + f'(x)\Delta x$$

This little formula gives us a means to approximate the values of functions at a variety of points.

When you get good at it, it makes for a great ice-breaker. Imagine being at a party where you spot a hunk of burning love across the room. You sidle over and casually say, "Do you happen to know $\sin(44°)$?" Your new friend whips out a calculator, but lo and behold, the battery is stone dead. Talk about egg on the face. But you say reassuringly, "No worries. Let me just borrow your napkin and a pencil. I can work it out."

Example *Find an approximation for* $\sin(44°)$.

Our function is $f(x) = \sin x$. So $f'(x) = \cos x$. Although we don't know $\sin x$ at 44°, we do know that $\sin 45° = \sqrt{2}/2$. So we'll pick $x = 45°$ and $\Delta x = -1°$.

But here is an *important point*. We should do all this in radians, not degrees. **Derivatives don't work out right when you use degrees.** So 45° becomes $\pi/4$ and $-1°$ becomes $-\pi/180$.

All right. Let's roll.

$$\sin(44°) = f(x + \Delta x) \approx f(x) + f'(x)\,\Delta x$$

$$= \sin\left(\frac{\pi}{4}\right) + \cos\left(\frac{\pi}{4}\right)\left(-\frac{\pi}{180}\right)$$

$$= \frac{\sqrt{2}}{2} + \frac{\sqrt{2}}{2}\left(-\frac{\pi}{180}\right)$$

$$= 0.694765$$

How good an approximation is that? Well, to eight decimal places, $\sin(44°) = 0.69465837$. So our approximation was accurate to the fourth decimal place. Amazing.

Knock off a few more approximations and your visually striking new friend will be following you around like a puppy dog. Of course, within a week, you will realize that judging people's worth by their complexions will mean that you spend a lot of time discussing the intricacies of the plot of *Baywatch*.

Intermediate Value Theorem and Mean Value Theorem

 20.1 Intermediate value theorem: It ain't a sandwich unless there's something between the bread

Okay, this theorem sounds nasty. It has a word with 12 letters in its title. On the other hand, as far as 12-letter words go, "intermediate" is not so awe-inspiring. Nor is this theorem. It really says precisely what its title claims, namely, that there are intermediate values. Here's an example. Suppose that you weighed 120 pounds when you were 15 years old and you now weigh 250 pounds at age 45. (It's probably not a good idea to draw a picture of this.) Then sometime between the ages of 15 and 45, you weighed 200 pounds, right? Because, somehow, you had to get from 120 to 250 pounds, and to do so you had to pass through 200 pounds. Perhaps you even remember the slice of chocolate pie that took you over? Anyway, that's all the **intermediate value theorem** says. It says that if you have a continuous function f on an interval $[a, b]$ (in our example the continuous function was your weight as a function of your age in years, where your age is in the closed interval $[15, 45]$) and if p is any value between $f(a)$ and $f(b)$ (in our case we took $p = 200$, which was a value between 120 and 250), then there must be a number c in the interval $[a, b]$ such that $f(c) = p$ (in our case there is some time between the ages of 15 and 45 where you weighed 200 pounds).

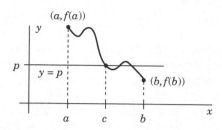

Figure 20.1 The graph must cross the line $y = p$ one or more times.

Graphically (see Figure 20.1), this theorem says that if f is continuous on $[a, b]$ and p is a value between $f(a)$ and $f(b)$, then the horizontal line at height $y = p$ will intersect the graph of $y = f(x)$ in at least one point with x coordinate between a and b. That point is the one we called c. There may be many points at which the curve crosses, but we can just pick one of them and call it c.

Although this theorem looks self-evidently true, it is relatively difficult to prove. Instructors in a first calculus course do not prove it unless they are from another planet, and we don't mean a nearby one.

So how do we use this theorem? Suppose that we want to know if $f(x) = x^4 - 7x^3 - 4x + 8$ is ever 0. Since this function is a polynomial, we know that it's continuous everywhere. At $x = -1$, we get $f(-1) = 20$. At $x = 1$, we get $f(1) = -2$. So at the two endpoints of the interval $[-1, 1]$ the function has values 20 and -2. Therefore, it must take on all values between -2 and 20 as x varies between -1 and 1. In particular, it must take on the value 0 for some x in $[-1, 1]$. The intermediate value theorem doesn't tell us exactly where it equals 0, but it does tell us that it is 0 somewhere on the interval $[-1, 1]$.

 Mean value theorem: Steep is steep

It was a slow day when the message came in. "Sam Spade's Calculus Agency," I said. It was the amusement park. They were stumped by the new ride they were building. It was going to be a wild new type of train ride, a "coaster" they were going to call it, starting and ending at ground level. What had their staff stumped was the design. They wanted to make sure that the customers would get those stomach-twisting effects when you shoot up, hang momentarily in the air, and plunge down. Or at least those bone-grinding moments when you drop like a rock, teeth jarring as you reach bottom, and launch upward with your appendix stretched down to your big toe. The manager was worried. Could she be sure the engineers wouldn't mess it up? After all, the water ride leaked, the tigers kept escaping, and as for the square ferris wheel—people seemed to like the kind that turned better.

"No problem," I assured her. "Rolle's theorem applies. No matter how they design it, the coaster will have at least one point where its slope is horizontal. Even your design crew can't go wrong. Now send me my fee, I'm a little short."

Well, the next summer I stopped to check out the new ride. A "Rolle Coaster," they called it. There wasn't much of a line though. The park engineers had made it completely flat. Instead of one or two points where the slope was horizontal, it was horizontal at EVERY point. "Should have told them about that case," I thought. Oh well, the check bounced anyway.

So just what does Rolle's theorem say? It says that if you start and end up at the same level, you will have a local max or min somewhere.

Theorem (Rolle's Theorem)   *If $f(x)$ is a continuous function differentiable on the interval $[a, b]$ and if $f(a) = 0$ and $f(b) = 0$, then there exists a point c in the interval $[a, b]$, where $f'(c) = 0$.*

In particular, this means that $f(x)$ has a peak where $f'(x) = 0$ or a valley where $f'(x) = 0$ or one more possibility: $f'(x) = 0$ everywhere, and the graph is flat at all places (see Figure 20.2). (In calculus, a point near which the graph is flat counts as both a local max and a local min! Talk about having it both ways.)

One fact to note about Rolle's theorem is that one can state it in the form:

There is a tangent line to the graph of $f(x)$ that is parallel to the line between the endpoints of the graph.

And now, on to one of the most well known of the theorems in calculus. This is a theorem that, upon entering the ballroom at the Calculus Cotillion, can get a hush to fall over the propositions, lemmas, and corollaries lining the walls. That's right, we are talking about the *mean value theorem*. As glamorous as it is, the surprising truth about the mean value theorem is that it is just a skewed version of the less ostentatious Rolle's theorem. Just view Figure 20.2 with your head tilted. You get Figure 20.3, which gives the mean value theorem.

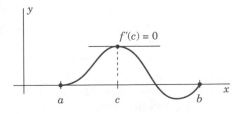

Figure 20.2 Rolle's theorem.

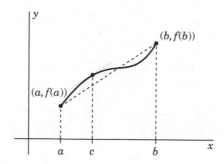

Figure 20.3 Mean value theorem.

Instead of assuming that $f(a) = 0 = f(b)$ as we did for Rolle's theorem, $f(a)$ and $f(b)$ can be anything. But the conclusion of the theorem will be the same, namely, *there is a tangent line to the graph of $f(x)$ that is parallel to the line between the endpoints of the graph.* Now we have to figure out how to say that mathematically. In this case, the endpoints of the graph are $(a, f(a))$ and $(b, f(b))$. So the slope of the line between these endpoints is $\dfrac{f(b) - f(a)}{b - a}$. We want there to be some tangent line to the graph with slope equal to this slope. That is to say, we want there to be some c in the interval $[a, b]$ such that $f'(c) = \dfrac{f(b) - f(a)}{b - a}$. So here is the theorem, which follows from Rolle's theorem (although the formal proof requires a little more work than tilting your head; you also have to jump up and down on one foot).

Theorem (Mean Value Theorem)   *Let $f(x)$ be a function that is continuous and differentiable on the interval $[a, b]$. Then there exists a point c in the interval $[a, b]$ where $f'(c) = \dfrac{f(b) - f(a)}{b - a}$.*

This theorem plays a part in the proof of many central results in calculus. It is important enough to have its own acronym, MVT. If your professor discusses the mean value theorem, pay heed, even if its relevance is not immediately apparent. If not, breathe a sigh of relief, and forget we ever brought it up.

Integration: Doing It All Backward

Having already discussed the first part of calculus, which revolves around the notion of differentiation, we are ready for the second part of calculus, which is based on the concept of integration. This is often confused with ingratiation, which is another calculus technique you'll want to learn to improve your grade. However, we'll stick to mathematical techniques in this chapter.

Integration is the undoing of differentiation, and of quite a few calculus students, too. But it needn't be. In this chapter, we'll outline the two basic varieties of integrals, cover some initial "techniques of integration," and end up with the "theoretical underpinnings." By the time we're done with this chapter, you will be an intoxicating interpretive integrator who commands adulation at math cocktail parties. If you don't go to math cocktail parties, expect an invitation in the mail shortly. We're desperate for friends.

There are two types of integrals in integration, the *definite integral* and the *indefinite integral*. They act much like the *definite person* and the *indefinite person* when you ask them for a date. The definite person answers "No! Drop dead! I'd rather do word problems about goats." The indefinite person says, "Well, I was planning to wash my hair on Friday unless it rains. Thursday my friend from Cleveland is coming to visit, but if there is a big storm and the airport closes, then I may be free. It depends on whether I finish the essay that's due tomorrow. I might be available Wednesday if I don't get a big assignment in my other class." In other words, the definite person gives you a specific and clear answer (in this case "No, you disgust me"). But the indefinite person returns a function, which can give either a yes or a no answer depending on a set of variables. In just the same way, the definite integral of a function produces a specific number, like 3 or 17. The indefinite integral of a function produces a function, like x^2 or $\sin x$. This is called the *antiderivative,* and it is found by doing the opposite of taking a derivative.

Let's take a more detailed look at the indefinite integral.

 ## Indefinite integral

An indefinite integral of a function is the antiderivative of the function. It is a new function whose derivative is the original function. We just reverse the process of taking derivatives, sort of like Bizarro differentiation.

Example   *Find the indefinite integral of $f(x) = 2x$.*

Choosing the function $f(x) = 2x$, we have that an indefinite integral (integral for short) of $f(x)$ is the function x^2, since

$$\frac{d}{dx}(x^2) = 2x$$

The function x^2 is an antiderivative of $f(x) = 2x$. In symbols, this is written

$$\int 2x \, dx = x^2$$

Some claim the integral symbol $\int$ has its origin in the letter "S," standing for "sorry about this old chaps," Newton's reported apology to his students for foisting the integral on them. Others claim it actually stands for "sum." By their reckoning, Newton was not the kind of guy to apologize for anything. The function inside the integral, in this case $2x$, is called the *integrand.*

Good Question   What is the little dx at the end of the integral?

Wrong answer: A 31-point word for Scrabble.
Right answer: An indication that x is the variable in the formula to be integrated. If you tried to figure out

$$\int 2 = ?$$

how would you know if you should put down $2x$ or $2t$ or $2y$? All three give 2 as a derivative:

$$\frac{d}{dx}2x = 2 \qquad \frac{d}{dt}2t = 2 \qquad \frac{d}{dy}2y = 2$$

But if you are told to calculate

$$\int 2\,dx$$

then you know you should write $2x$ and not $2t$.

Let's look again at the formula

$$\int 2\,dx = 2x$$

The function $2x$ is not the only one whose derivative equals 2. Another is $2x + 3$, since

$$\frac{d}{dx}(2x + 3) = 2$$

So are $2x + 7.134$ and $2x - 5$. In fact any constant can be added to $2x$ without changing its derivative. In order to include all these possibilities, we write

$$\int 2\,dx = 2x + C$$

where C represents any constant.

Remember for exam: *Don't forget the constant!*

You are almost certain to lose a point or two on some exam by forgetting to put $+C$ after calculating an indefinite integral. This is a standard rite of passage in the calculus learning process.

+C Joke   Two math professors go into a bar, arguing about how much the average person knows about calculus. As one sits down at a table, the other goes to the bar and orders drinks from the bartender. She then says to the bartender, "When you bring the drinks over, I'll ask you a question, and whatever it is, you answer, "x to the fourth over four." Then she slips him a fiver.

When she returns to the table, she says to her friend, "I think the average joe knows more about calculus than you think. In fact, I will bet you ten dollars that if we ask the bartender the integral of x^3, he'll know the answer." "Hah," says the friend. "You're on." When the bartender arrives at the table, the first professor says, "Excuse me, but can you answer a question for us? What's the integral of x^3?" The bartender answers, "That's easy, x to the fourth over four ... $+C$."

The purpose of this joke was not to make you laugh, although when told at a math event, professors have been known to laugh so hard they shoot Sprite out their noses. No, the purpose of the joke is to remind you to always include the "$+C$" as part of your answer whenever you take an indefinite integral.

Bad Question   Why do we always have to put that stupid $+C$ on the end of the answer to an indefinite integral?

It's always a bad idea to call things "stupid" that others hold dear. You probably wouldn't say to the professor, "I met your son yesterday, and he's stupid." A more diplomatic question might be "Does the $+C$ serve as a reminder that there are an infinite number of functions that are the indefinite integrals of a function, each corresponding to a different choice of value for the C?"

21.2 Integration method: The easy ones

Since integration is the reverse of differentiation, we can reverse the rules for differentiating to get integration rules. How about the power rule for a start?

$$\frac{d}{dx}x^n = nx^{n-1}$$

Running it backward, we get what we can call the **power rule for integration:**

$$\boxed{\int x^n\ dx = \frac{x^{n+1}}{n+1} + C}$$

This is a reasonable answer, since if we reverse the process by taking the derivative of our answer, we obtain

$$\frac{d}{dx}\left(\frac{x^{n+1}}{n+1} + C\right) = \frac{(n+1)x^n}{n+1} = x^n$$

So $\dfrac{x^{n+1}}{n+1} + C$ *is* the antiderivative of x^n!

To use this rule of integration, you must, without fail, be able to add 1 to any number. Note that the inverse power rule does have a problem when $n = -1$. Then we get a 0 in the denominator of our answer, a sin in mathematical circles of such great magnitude that they take away your pocket protector. So, *the power rule for integration does not apply to integrate* $1/x$.

The right answer in that case is

$$\int \frac{1}{x}\, dx = \ln|x| + C$$

Sorry about that, but that's the way it is. Learn to live with it. We will discuss this some more in Chapter 24.

Example *Calculate*

$$\int x^3\, dx$$

Here $n = 3$, so

$$\int x^3\, dx = \frac{x^4}{4} + C$$

Example *Calculate*

$$\int (2x^{-1/3} + 3x^3)\, dx$$

Do this piece by piece, carrying along constants as you go:

$$\int (2x^{-1/3} + 3x^3)\, dx = 2\frac{x^{2/3}}{2/3} + 3\frac{x^4}{4} + C$$

$$= 3x^{2/3} + \frac{3x^4}{4} + C$$

The last example used the facts, explanations of which weigh in at about 2.2 pounds in your standard calculus book, that when you multiply a function by a constant (like 2) and when you add two functions together, the anti-derivatives behave themselves, just like the derivatives did. (This is as opposed to what happens when you multiply two functions together, when all hell breaks loose.)

Anything else we can easily integrate?

Well, since $\dfrac{d}{dx}(\sin x) = \cos x$, we know that

$$\int \cos x \, dx = \sin x + C$$

Similarly,

$$\int \sin x \, dx = -\cos x + C$$

It's even true that

$$\int \sec^2 x \, dx = \tan x + C$$

since

$$\frac{d}{dx}(\tan x) = \sec^2 x$$

The Tale of the Pit and the Goat   What is the indefinite integral good for? Well, here's a story about a time when it came in useful.

Two animal rights activists were walking in a field when they came across a huge dark hole in the ground. They leaned over the edge and peered into the blackness. "How deep do you think it is?" said the woman. "Beats me," said the man. "But hey, let's throw this rock in and count how long it takes until we hear it hit bottom." So together, they hoisted up the rock and heaved it over the side. As they counted to keep track of the time, they waited to hear the rock hit bottom. Finally, they heard what might have been a distant splash. "Was that it?" said the woman. "I don't know," said the man. "We need something bigger." The woman glanced around the field and spotted an old railroad tie. "Look," she said. "That ought to do the trick." With much effort, the two of them managed to haul the tie over to the edge of the hole and heave it over the side, where it disappeared into the blackness. They counted up to 6 seconds, when suddenly a goat raced up to the edge of the hole and dove headfirst over the side. Momentarily stunned, they continued to count, hearing first a big splash in the water below at 10 seconds and a smaller splash shortly afterward.

At that instant a farmer yelled from the edge of the woods bordering the field. "Hey, have you seen my goat?" "Yes," the woman yelled back. "A goat just ran up here and jumped in this hole." "Oh, that couldn't have been my goat," the farmer yelled back. "My goat is tied to a railroad tie."

Question: How deep was the hole?

We know that the tie went over the side at time $t = 0$ and splashed in to the water below at time $t = 10$. We also know that gravity imparts an acceleration of $a(t) = -32$ ft/sec^2 to any object falling near the surface of the earth, animal, vegetable, or mineral. We will ignore the animal and focus on the vegetable, which is to say, the railroad tie. Since the weight of the goat is relatively small in comparison to the weight of the tie, we will disregard the drag caused by the goat. Since the derivative of velocity is acceleration, we know that the indefinite integral of acceleration is velocity. In symbols, that's

$$v(t) = \int a(t)\,dt$$

So

$$v(t) = \int -32\,dt = -32t + C_1$$

Since the velocity of the tie at time $t = 0$ was 0, we know that $v(0) = 0$. This means that $C_1 = 0$, since $0 = v(0) = -32(0) + C_1$.

Now for the position function:

$$s(t) = \int v(t)\,dt = \int -32t\,dt = -16t^2 + C_2$$

The height of the railroad tie at time $t = 0$ was also 0. So $s(0) = 0$ and therefore $C_2 = 0$ too. The equation for the height of the railroad tie after t seconds is now pinned down exactly, with no more constants undetermined.

$$s(t) = -16t^2$$

By plugging in $t = 10$ we can see that the depth of the hole is $s(10) = -16(100)$. The hole is 1600 feet deep, the minus sign indicating that it is below, rather than above, ground level.

P.S. The goat never had a chance.

 21.3 Integration method: Substitution

There you are coaching the big game, and your team, the Variables, is losing bigtime. It's not clear what went wrong, but the Integrals are tearing you apart. Your star player x is just not performing. He can't seem to get going. You hate to do it, but you have no choice: Time to send in the sub. You look down the bench, and every one of of your players is staring down at the floor, afraid to catch your eye. They don't want to go out there just to be humiliated. But way down at the end, that skinny kid u is looking back at you hopefully. "That kid's got spunk," you think to yourself. "Well, what the hell, this game's lost anyway." You yell out, "Okay, x, you're out of there, hit the bench." Then you point down at u. "All right, kid. This is your big chance. I'm putting you in for x." You turn to the referee and utter those fateful words, "I'm making a u substitution . . ."

Hey, wake up, dream's over . . . remember, calculus, getting that A? In this section, we will show you one of the most powerful techniques for finding integrals. It is the reverse of the chain rule, one of the most powerful rules for differentiation. You remember the chain rule:

$$(f(g(x)))' = f'(g(x))g'(x)$$

What do we get when we reverse it to do integration?

Substitution $\displaystyle\int f'(g(x))g'(x)\,dx = \int (f(g(x)))'\,dx$

$$= f(g(x)) + C$$

This is usually written with a u instead of a g, in the following equivalent way:

$$\int f'(u(x))u'(x)\,dx = \int (f(u(x)))'\,dx$$

$$= f(u(x)) + C$$

If we write

$$u'(x)\,dx = \frac{du}{dx}\,dx = du$$

then the equation becomes

$$\int f'(u(x))\,du = f(u(x)) + C$$

We seem to have done something very naughty here. We canceled dx's and replaced $\dfrac{du}{dx}\,dx$ by du. This is okay, though. The whole notation has been set up to allow us to treat $\dfrac{du}{dx}$ as a fraction, without getting into trouble. Notation is power!

In practice, nobody bothers to remember the formula. Instead, just remember to turn all the parts of the integral involving x, including the dx, into parts involving u. The hard part is deciding what part of the integrand should be equal to u. Time for some examples.

Example   *Calculate $\int \sin^2 x \cos x \, dx$*

Pick u to be $\sin x$. Then

$$u(x) = \sin x$$

$$\frac{du}{dx} = \cos x$$

so

$$du = \cos x \, dx$$

The advantage of this maneuver is that you can replace the wretched $\cos x \, dx$ in your integral with du, and the $\sin x$ with u, and the integral becomes the easy

$$\int u^2 \, du = \frac{u^3}{3} + C$$

This trick of multiplying through by dx seems a little sleazy, since $\dfrac{du}{dx}$ is not an actual fraction, but rather a notation for the derivative. As any law firm will attest, "Sure it's sleazy, but it's legal, so go ahead and enjoy!"

But you're not finished yet. It doesn't make any sense to start off with a question about a function of x and end up with an answer about a function of u. It's like answering "helicopter" when someone asks you what time it is.

Rewrite the answer in terms of x by replacing u with $\sin x$:

$$\tfrac{1}{3} \sin^3 x + C$$

Example   *What is $\int e^{x^3} x^2 \, dx$?*

Pick $u = x^3$. Then

$$\frac{du}{dx} = 3x^2$$

$$du = 3x^2\,dx$$

Don't let a constant throw you—just divide both sides by 3:

$$\tfrac{1}{3}\,du = x^2\,dx$$

The integral is just:

$$\int e^{x^3}x^2\,dx = \int e^u \frac{1}{3}\,du = \frac{1}{3}\int e^u\,du$$

$$= \frac{e^u}{3} + C$$

$$= \frac{e^{x^3}}{3} + C$$

21.4 Integration method: Eyeball technique

Perhaps the most commonly used method of integration, the eyeball technique rarely appears in textbooks. The idea is simple. We eyeball the integral, take a guess at the answer, and differentiate it to see if we were right. If not, we can often see what went wrong and adjust our guess accordingly. So, for example, say we want to find $\int (2x + 1)^4\,dx$. We could certainly do this by substitution, but let's pretend, just for an instant, that we are lazy and just don't want to be bothered. So we guess that it's $(2x + 1)^5/5 + C$. Well, we can check our guess by differentiating, using the chain rule.

$$\frac{d}{dx}\left[\frac{(2x + 1)^5}{5} + C\right] = \frac{5(2x + 1)^4(2)}{5} = 2(2x + 1)^4$$

The derivative of our guess differs from the original integrand by a factor of 2. It has an extra 2 in it. So if we adjust our guess to be

$$\frac{(2x + 1)^5}{10} + C$$

then we won't end up with that extra 2 in the derivative and therefore this gives the correct integral. Pretty easy, huh? But always fall back on substitution if there is any confusion. The eyeball technique should be used only when you are feeling pretty cocky.

21.5 Integration method: Tables

Any self-respecting calculus book has a table of indefinite integrals that either appears as an appendix to the book or else appears on the inside of the front and back covers of the book. If your calculus text does not contain such a table, immediately write a nasty letter to the publisher. It won't get you a table, but you'll feel a lot better. These tables have the advantage that if you can't figure out how to do a particular integral, you can flip to the table, find your particular integral, and write down the answer. Sounds easy enough. But there are three caveats:

1. If you are doing homework to learn a particular method of integration, like substitution, you will probably get no credit for writing down an answer that you found in the table.

2. It is easy for an instructor to construct an integral that doesn't appear in any table. So it just may not be there.

3. Even if your integral is in the table, it may be in disguise. You may have to manipulate the integral to get it in the right form.

Example *Find*

$$\int \frac{1}{x^2 - x - 2}\, dx$$

You eagerly flip to the table only to find nothing that looks like this. Ah, but

$$\frac{1}{x^2 - x - 2} = \frac{1}{(x - 2)(x + 1)}$$

and there, sitting as pretty as you please in the table's formulae (you gotta love Latin plurals), is

$$\int \frac{1}{(x - a)(x - b)}\, dx = \frac{1}{a - b}(\ln|x - a| - \ln|x - b|)$$

So letting $a = 2$ and $b = -1$, we have

$$\int \frac{1}{x^2 - x - 2}\, dx = \int \frac{1}{(x-2)(x+1)}$$

$$= \frac{1}{2-(-1)}\left(\ln|x-2| - \ln|x-(-1)|\right)$$

$$= \tfrac{1}{3}\left(\ln|x-2| - \ln|x+1|\right)$$

What exactly is this "ln" function? That's coming up in the next chapter. Something to look forward to.

21.6 Integration method: Computers and calculators

Yes, it's true. There are calculators available for $50 that can do any integral in your calculus class. Programs such as *Maple, Mathcad, Mathematica,* and *DERIVE* are even better, able to do integrals that no rational person would attempt. So why not just use these and skip the other methods?

There are many possible answers to this question, but the fundamental reason is that if you use these tools without a basic understanding of the process of integration, you're likely to use them incorrectly. Incorrect calculations of integrals can lead to collapsing bridges, inefficient allocation of financial resources, unwanted pregnancies, poor ticket sales, excessive saltiness, and other problems we are much too polite to mention.

The Definite Integral

22.1 How to find the definite integral

The definite integral starts with a function and a couple of numbers and produces a (definite) number. A definite integral looks a lot like an indefinite integral, only it has little numbers at the top and bottom of the integral symbol. These are called the *limits of integration*. So if someone asks you, "What are your limits?" the correct response may be "−1 and 2" rather than "Well, I refuse to crawl through Jell-O in my underwear. That I will not do."

So here is how we write a definite integral:

$$\int_0^4 x \, dx$$

How do we compute it? Easy! First, find the indefinite integral (that is, the antiderivative) of x. That would be $x^2/2 + C$. Then erase the C. (We know, it makes the C seem sort of pointless, doesn't it?) Then plug the top limit "4" into $x^2/2$, giving $4^2/2 = 8$, and subtract from that $x^2/2$ with the lower limit "0" plugged in.

To write this all in one line, we would say

$$\int_0^4 x \, dx = \frac{x^2}{2}\bigg|_0^4 = \frac{4^2}{2} - \frac{0^2}{2} = 8$$

So the answer is 8. Pretty easy.

Why did we dump the $+C$? Since we take the function evaluated at the top limit and subtract from it the function evaluated at the bottom limit, the C just cancels out anyway, so why bother keeping it around? If you are not going to eat it, don't put it on your plate.

Let's do another one, a little harder.

$$\int_0^\pi \sin x \, dx = -\cos x \bigg|_0^\pi = -\cos \pi - (-\cos 0) = -(-1) - (-1) = 2$$

Of course, we have yet to discuss the most important aspect of definite integrals, namely: So what? Who cares? What's the point? Put more politely, why should we ever want to compute one? What are they telling us?

In a word, AREA.

 ## 22.2 Area

That's right. A definite integral is computing an area for you. What area? Well, if you have the integral

$$\int_a^b f(x) \, dx$$

then it is giving you the area under the curve $y = f(x)$ (and above the x axis) between the values of $x = a$ and $x = b$ (see Figure 22.1).

So, for instance,

$$\int_0^4 x \, dx$$

should give the area under the curve $y = x$ from $x = 0$ to $x = 4$. If we look at the shaded area in Figure 22.2, we see that it's the area of a right triangle of height 4 and base 4, which is equal to 8, and that is exactly what we got for the definite integral. Good.

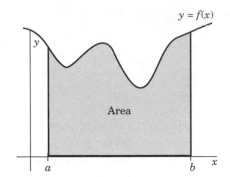

Figure 22.1 The definite integral gives the area under the graph between $x = a$ and $x = b$.

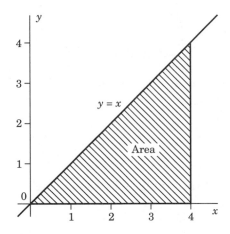

Figure 22.2 Area under a right triangle.

Of course, we didn't need a definite integral to figure out the area of a triangle. On the other hand, since

$$\int_0^\pi \sin x \, dx = 2$$

we know that the area under the sine curve from $x = 0$ to $x = \pi$ in Figure 22.3 is exactly 2, something we would have had a hard time figuring out without integrals.

Great, this whole integral thing seems pretty straightforward. But now that the waters of understanding are calm and clear, it's time to run through

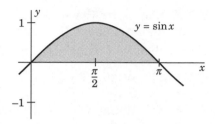

Figure 22.3 The area under the sine curve between 0 and π is 2.

in a speedboat, muddying the waters and endangering any manatees floating nearby. A definite integral

$$\int_a^b f(x)\,dx$$

does not necessarily always give the area under the graph of the function between $x = a$ and $x = b$. We sort of lied. But as with most good lies, there is a core of truth.

A truer fact is that if the graph of the function $f(x)$ always stays above the x-axis for x between a and b, then the definite integral gives the area under the graph. However, if the graph of $f(x)$ dips below the x-axis, the definite integral gives us the the area above the x-axis minus the area below the x-axis. See Figure 22.4.

So for example,

$$\int_0^{2\pi} \sin x\,dx = 0$$

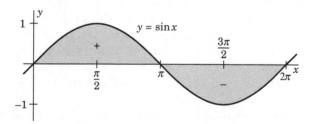

Figure 22.4 The definite integral can be negative if the curve dips below the x-axis. The integral from 0 to 2π of $\sin x$ equals zero!

either by direct calculation or by the fact that the area under the x-axis is exactly the same as the area above the x-axis, so the two will cancel in calculating the definite integral, giving an answer of 0.

FABLE OF A KING AND A GOAT

Once upon a time there was a king who lived alone in his drafty castle with his three lovely and intelligent daughters and their pet goat. His daughters had reached the age where it was time to meet that special someone, but up to now, all the men who interested them were bikers and no-good rogues. So the king decided to devise a test as a challenge to their suitors, figuring that bikers wouldn't stand a chance. He announced to all his subjects that anyone who could tell him exactly how many peasants lived in the kingdom would receive 1000 gold pieces and the choice of one of his daughters as his bride. Anyone who guessed wrong would lose his head.

Now the king knew that this was a difficult problem, because although the density of peasants was legislated to be exactly $15/8$ peasants per square mile, the kingdom had an irregular shape. Three of its borders were straight, one of length 100 miles, one of length 110 miles, and the third of length 10 miles, but the fourth side was bounded by a curving river, making a calculation of the area appear impossible. Many tried, though, lured by the shining prizes. Within a year, there were no more bikers in the kingdom, much to the king's satisfaction and the princesses' despair. "C'mon, Dad, quit screwing around like this," they would say. "Get a life. No one around here can even differentiate. We'll be single forever."

Then one day a timid young man with a pocket protector came to the king and said, "I have come for the gold and to marry one of your daughters." The king laughed. "Is that so? Then tell me the number of peasants in the kingdom."

"8,124.5," said the young man. The king's jaw dropped a foot. What sorcery was this? It was amazing enough that he had gotten the whole number right, but how did he know about the half-time peasant who turned into a werewolf for two weeks each month?

What the king didn't realize was that the timid-looking young man was an itinerant calculus professor, touring the kingdom on his bicycle. He had heard about the king's challenge and realized it was his best chance of ever getting a date. Having just ridden his bike along the river, he knew that its course was given by the curve $y = x^2/100 + 10$. A map of the kingdom would therefore appear as a rectangle of width 100 but with a curved top (see Figure 22.5).

Thence, therefore, and thereby, the area of the kingdom was given by the definite integral

$$\int_0^{100} \frac{x^2}{100} + 10 \, dx$$

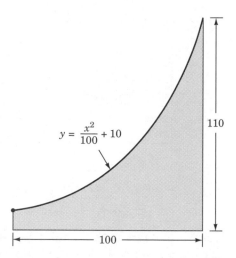

$$y = \frac{x^2}{100} + 10$$

Figure 22.5 The kingdom.

This integrated to $\left. \dfrac{x^3}{300} + 10x \right|_0^{100} = 4333.333\ldots$ Then, multiplying by the peasant density of $^{15}\!/_8$ of a peasant per square mile, he obtained the total peasant population, $(4333.333\ldots)(^{15}\!/_8) = 8,124.5$.

The king, good to his word, immediately sent for the court chamberlain to perform the marriage ceremony. "Go to the royal chambers," he ordered the young man, "and bring out the bride of your choice." The timid young man soon emerged with his bride to be, beaming triumphantly. The wedding ceremony was quickly performed, and he picked up his sack of gold and headed for the door with his new spouse.

"Wait," said the king. "If I can guess your profession, will you return my 1000 pieces of gold?" The timid young man realized that he had never said anything about himself or his career, and the king could have no idea what he did for a living. Basking in his victory, he experienced a rush of magnanimity, and said, "Sure, give it your best shot." The king smiled, "I would say you're a calculus professor." The young man's jaw dropped 2 feet. "How did you know?" "Well," said the king. "You just married my goat."

Incidentally, this story does have a politically correct ending. When the rabble heard about the shabby way that the calculus professor was treated, they stormed the castle and deposed the king, freeing his daughters from the shackles of castle life. All three of the daughters offered their undying affection to the calculus professor who had been the catalyst for freedom from their father's tyranny, but he spurned them, remaining faithful to his one true love. So they all turned to careers in mathematics, and became happy

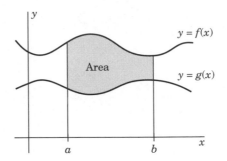

Figure 22.6 Area between $y = f(x)$ and $y = g(x)$ above the interval $[a, b]$.

successful university professors with big fat salaries. (Hey, it's our fairy tale, if we say they have big fat salaries, then they do.) The End.

P.S. The goat never had a chance.

At this point, we can now find the area under a curve. Put another way, we can find the area between a curve and the x axis. Of course, how often are we interested in finding the area under a curve? How often have you said to your friend, "Hey, it's Friday night. Let's go find some areas under curves?"

But finding areas between curves, now that's excitement. That's action! That is a different kettle of fish entirely. How do we do this?

Suppose that we have two graphs, one for the function $f(x)$ and one for the function $g(x)$ (see Figure 22.6). Suppose, for convenience, that the graph of $f(x)$ is always above the graph of $g(x)$. [So $f(x) > g(x)$ for all x.] Then the area under the graph of $f(x)$ between $x = a$ and $x = b$ is given by $\int_a^b f(x)\,dx$ and the area under the graph of $g(x)$ between $x = a$ and $x = b$ is given by $\int_a^b g(x)\,dx$. So the area between $f(x)$ and $g(x)$ is just

$$A = \int_a^b f(x)\,dx - \int_a^b g(x)\,dx = \int_a^b (f(x) - g(x))\,dx$$

Stating this carefully: If $f(x) \geq g(x)$ for all $x \in [a, b]$, then the area between $y = f(x)$ and $y = g(x)$ over the interval $[a, b]$ is given by

$$A = \int_a^b (f(x) - g(x))\,dx$$

Example   *Find the area between $y = x^2$ and $y = x^3$.*

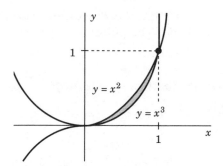

Figure 22.7 Area between $y = x^2$ and $y = x^3$.

This is a famous example that appears in almost every calculus book as either an example or a problem.

You will notice from Figure 22.7 that the problem is really asking for the area of the crescent moon region shaded in the figure, since that is the only region bounded by the two curves that will have a finite area. In order to find its area, we first have to find the intersection points of the two curves. Setting $x^2 = x^3$, we obtain

$$x^2 = x^3$$

$$x^2 - x^3 = 0$$

$$x^2(1 - x) = 0$$

$$x = 0 \qquad \text{or} \qquad x = 1$$

The second fact we need to note is that when x is between 0 and 1, x^2 is *bigger* than x^3. So the top function is $y = x^2$ and the bottom function is $y = x^3$.

Then the formula for areas tells us that

$$\text{Area} = \int_0^1 (x^2 - x^3)\, dx = \left. \frac{x^3}{3} - \frac{x^4}{4} \right|_0^1 = \frac{1}{3} - \frac{1}{4} = \frac{1}{12}$$

22.3 Fundamental theorem of calculus

We have yet to discuss why definite integrals and the quantities that they represent, such as areas, can be obtained in this way. The answer is in that crown jewel of mathematics, the **Fundamental Theorem of Calculus.** It tells us that our method of computation is correct.

Theorem (Fundamental Theorem of Calculus) $\displaystyle\int_a^b f(x)\,dx = F(b) - F(a)$

where $F(x)$ is an indefinite integral (or antiderivative) of $f(x)$.

In this equation the left-hand sign is defined to be something like the area under a curve, if the curve is positive, or more accurately something called the *Riemann sum* which we'll get to shortly. So the Fundamental Theorem says there is an easy way to calculate areas and Riemann sums. Sounds pretty nifty when you think of it like that.

Why are we stating it in this much detail when it is already apparent how to apply it? One good reason is that there are professors out there, and we won't name names, who believe that a good exam question is: "State the Fundamental Theorem of Calculus." To do this precisely, you will need to add on some conditions about continuity of $f(x)$, or perhaps integrability. Check your notes for these.

The point of the Fundamental Theorem is to take something you want to know (the left-hand side of the equation) and translate it into something you have a hope of calculating (the right-hand side). We didn't say how *much* hope you have of calculating the right-hand side, just that you have some. If you can't find an antiderivative of $f(x)$, you're sunk. This means that practically speaking, in the real world, you're sometimes out of luck. Functions just don't always have nice antiderivatives. However, in the carefree happy-go-lucky world of first-year calculus, you are asked to calculate integrals of functions with (relatively) easy-to-find antiderivatives. So almost all of integral calculus (the bit we're doing now) is dedicated to *techniques of integration,* that is, finding antiderivatives. The idea is that you will develop a collection of techniques which will work in most situations, though maybe not all.

We also want to mention another theorem which often goes by the title of the Fundamental Theorem of Calculus Part II, or sometimes the Fundamental Theorem of Calculus Part I, making our previous fact the Fundamental Theorem of Calculus Part II. It states the famous relationship between integrals and derivatives explicitly, namely, that they are inverses of one another. But it does this for derivatives and definite integrals. Here it is:

$$\frac{d}{dx}\int_a^x f(t)\,dt = f(x)$$

In words, let's see what this says. Start with a function $f(t)$. Apply an integral to it, in this case with one limit a that is a constant, and one limit x that is a variable. The result will be some function of x. Now, differentiate that function of x. Lo and behold, the derivative cancels out the effect of the integral, and all we are left with is the original function f, but now as a function of x.

The one strange phenomenon here is that the result doesn't depend in any way on what you put in for the lower limit of integration a. You could put 4 in there, or π in there, or a dirty sock in there, the result would be the same. Besides giving an explicit meaning to the fact that integration and differentiation are inverse operations, this can be used to prove the fundamental theorem of calculus, and that is why it gets covered.

22.4 Some basic rules for definite integrals

Okay, here are some rules for playing around with definite integrals. They come in handy in a variety of situations. It's always a good idea to have a few tools in the old tool belt.

1.
$$\int_a^b f(x)\, dx = -\int_b^a f(x)\, dx$$

Switch the limits, switch the sign.

2.
$$\int_a^b cf(x)\, dx = c\int_b^a f(x)\, dx$$

You can pull constants out of integrals, just as you pull constants out of derivatives.

3.
$$\int_a^c f(x)\, dx = \int_a^b f(x)\, dx + \int_b^c f(x)\, dx$$

If we think of the integral as corresponding to an area under a curve, this becomes apparent. The area under $f(x)$ between points a and c should equal the area between a and b plus the area between b and c.

Example *Calculate* $\int_{-2}^1 |x|\, dx.$

Well,

$$|x| = \begin{cases} x & \text{if } x \geq 0 \\ -x & \text{if } x < 0 \end{cases}$$

So if we split the integral up, we can write

$$\int_{-2}^{1} |x|\, dx = \int_{-2}^{0} |x|\, dx + \int_{0}^{1} |x|\, dx$$

$$= \int_{-2}^{0} -x\, dx + \int_{0}^{1} x\, dx$$

$$= -\frac{x^2}{2}\Big|_{-2}^{0} + \frac{x^2}{2}\Big|_{0}^{1}$$

$$= (0 - (-4/2)) + (1/2 - 0)$$

$$= 5/2$$

22.5 Integration method: Numerical approximation

Numerical methods can just about always find a definite integral, at least approximately. They won't find an antiderivative function into which you can plug the limits to find the exact answer. Instead they just turn out a number which is pretty close to the definite integral. The idea is something like measuring the amount of Rocky Road ice cream in a bowl by counting how many spoonfuls it takes to eat it. Not nearly as elegant as a theoretical calculation giving the precise answer, but tasty nonetheless and accurate enough for the purpose.

For the many functions for which antiderivatives simply cannot be found, numerical integration is the way to go. A simple example of such is

$$\int_{0}^{1} \sin(x^2)\, dx$$

As tempting as it is to try to find a really tricky way of determining an antiderivative, resist the urge to waste your life on this. Nothing works, not substitution, not eyeballing, nothing. So just don't go there. Approximations are all you can hope for.

The main numerical methods involve approximating the area under a curve by rectangles (the rectangle rule or midpoint rule), trapezoids (the trapezoid rule), or pieces of parabolas (Simpson's rule—this has nothing to do with Bart Simpson; he uses hyperbolas). The more rectangles or trapezoids that are used, the more accurate the calculation. Computers seem to be better at this sort of thing than people, since computers would rather compute large sums than watch TV. Many calculators and computers have these types of methods built into them, and this is what they use when you ask them to find a definite integral for you.

Here's how all the approximation methods work. Suppose you're trying to approximate

$$\int_a^b f(x)\,dx$$

First cut the interval $[a,b]$ into n equal pieces, called subintervals. Each piece will have length $\dfrac{b-a}{n}$. Label the endpoints of the subintervals, going from x_0 to x_n. If you cut your interval into three pieces, say, then you need four labels to cover all the endpoints. And a and b get new labels, $a = x_0$ and $b = x_n$.

Each of the three numerical methods replaces the section of the graph of $f(x)$ over each subinterval with a different graph, which is pretty close to $f(x)$, but for which the area under the graph is easier to calculate.

If you replace bits of $f(x)$ with bits of horizontal lines, you get a whole set of rectangles that just about cover the area we are interested in (see Figure 22.8 on the following page).

The sum of the areas of the rectangles is approximately the area under the curve $y = f(x)$ from $x = a$ to $x = b$. Since each rectangle will have length $\dfrac{b-a}{n}$ and height $f(x_i)$, which is the height of the function at the left-hand endpoint, the area of the rectangle is $\dfrac{b-a}{n}f(x_i)$.

Adding up the areas of these rectangles, we get the rectangle rule.

Rectangle Rule (Using Left-Hand Endpoints)

$$\int_a^b f(x)\,dx \approx \frac{b-a}{n}[f(x_0) + f(x_1) + \cdots + f(x_{n-1})]$$

Notice you should stop at x_{n-1}, not x_n.

If we use the right-hand endpoints instead, we get a slightly different rectangle rule.

$$\int_a^b f(x)\,dx \approx \frac{b-a}{n}[f(x_1) + f(x_2) + \ldots + f(x_n)]$$

These are also called *Riemann sums,* after Georg Friedrich Bernhard Riemann, the guy who invented this idea. Even after dropping the "e" in George, his name was still so long that he had to break it up into little pieces just to remember it.

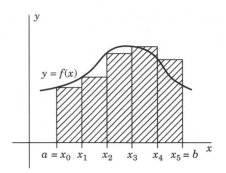

Figure 22.8 Approximating the definite integral $\int_a^b f(x)\,dx$ with rectangles.

If we choose rectangles whose height is given by the value of the function at the center of the rectangle, instead of the left-hand endpoint, we get an approximation of the integral called the midpoint rule (see Figure 22.9).

Midpoint Rule

$$\int_a^b f(x)\,dx \approx \frac{b-a}{n}\left[f\left(\frac{x_0+x_1}{2}\right)+f\left(\frac{x_1+x_2}{2}\right)+\cdots+f\left(\frac{x_{n-1}+x_n}{2}\right)\right]$$

where

$$\frac{x_0+x_1}{2},\qquad \frac{x_1+x_2}{2},\qquad \ldots,\qquad \frac{x_{n-1}+x_n}{2}$$

are the midpoints of the n equal intervals between $a = x_0$ and $b = x_n$.

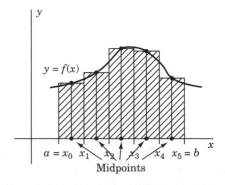

Figure 22.9 Midpoint rule.

This basic idea, approximating the area under a curve by the sum of the areas of rectangles, is one of the most important in all of math, right up there with approximating a tangent line with secant lines. We will come back to it in a second. But meanwhile...

If we replace bits of f with slanted straight lines that look a little more like $f(x)$ than the top of a rectangle, we get the trapezoid rule (see Figure 22.10).

Trapezoid Rule

$$\int_a^b f(x)\,dx \approx \frac{b-a}{n}\left[\frac{f(x_0)}{2} + f(x_1) + \cdots + f(x_{n-1}) + \frac{f(x_n)}{2}\right]$$

Here you include both $f(x_0)$ and $f(x_n)$, but only half of each. This comes out of the fact that the area of a trapezoid with left height L, right height R, and width W is $\dfrac{(L+R)W}{2}$.

If instead we replace bits of $f(x)$ with pieces of parabolas that look even more like $f(x)$, we get Simpson's rule.

Simpson's Rule

$$\int_a^b f(x)\,dx$$
$$\approx \frac{b-a}{3n}[f(x_0) + 4f(x_1) + 2f(x_2) + \cdots + 2f(x_{n-2}) + 4f(x_{n-1}) + f(x_n)]$$

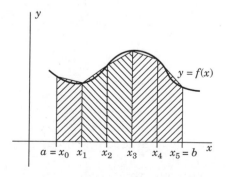

Figure 22.10 In the trapezoid rule, guess what we use instead of rectangles?

Watch out for the $(b-a)/3n$ in this rule. That's what comes of integrating parabolas. The pattern of coefficients for the $f(x_i)$'s is

$$1, 4, 2, 4, 2, 4, \ldots, 2, 4, 2, 4, 1$$

This works out as long as you cut $[a, b]$ into an even number of pieces. Your instructor may or may not go through an argument as to why the coefficients work out to be what they are. If it is covered in class, make sure you know it unless told explicitly that it will not be on the exam. It is usually considered a bit too messy for an exam problem.

These approximation rules are all straightforward to calculate but rather tedious. Just the thing for a computer or calculator, and in some classes, a calculus student. Programmable calculators and computers are the best way to go here. One can create a loop in the program that adds on a term in the sum each time it iterates, so that by the time the program finishes looping, you have the whole sum.

As an example, let's see how we would outline a program that would find an approximation to the integral $\int_1^2 \sqrt{x}\, dx$ using the rectangle rule (with the right endpoint) and 10 terms in the sum.

Sample Program for Rectangle Rule

```
1 S = 0
2 f(x) = sqrt(x)
3 a = 1
4 b = 2
5 n = 10
6 d = (b - a) / n
7 FOR k = 1 TO n
8 S = S + f(a + k * d) * d
9 NEXT k
10 PRINT S
```

The term f(a + k * d) * d in line 8 is the area of each subsequent rectangle, which gets added to the total area S as we repeatedly loop through the FOR loop in the program.

If we wanted a program using the left endpoint, we would replace the k in line 8 with (k − 1). If we wanted the midpoint rule, we would replace the k in line 8 with (k − 0.5), since 0.5d is half the width of the subintervals.

As far as exams go, the best approach to these approximation rules is to memorize the formulas right before the exam, if required, and forget them again right afterward.

22.6 Riemann sum—with nitty gritty details

Occasionally an instructor will demand that you know the technical definition of the definite integral. This is tricky at first exposure, although when you get the hang of it, it's not too bad. It's really just the rectangle rule for approximating an integral.

Remember the rectangle rule with the height of each rectangle corresponding to the value of the function at the right-hand endpoint? It says: Divide the interval $[a, b]$ into n segments $[x_0, x_1], [x_1, x_2], \ldots, [x_{n-1}, x_n]$, of equal length, where $x_0 = a$ and $x_n = b$. Then,

$$\int_a^b f(x)\, dx \approx \frac{b-a}{n}[f(x_1) + f(x_2) + \cdots + f(x_n)]$$

To figure out what

$$\int_a^b f(x)\, dx$$

is *exactly*, we have to make that approximation better and better. That translates into cutting the interval $[a, b]$ into smaller and smaller pieces, which means that the number of rectangles gets bigger and bigger. We get the exact value if you take the limit as the number of rectangles n goes to infinity. (Hey, limits again. Who would have thought?)

Tale of Two Camps   It was the best of times, it was the worst of times. Applications were up, but the collapse of their buildings in the winter storms meant that this summer the adolescent boys of Camp Wantaugh and the girls of Camp Peak would have to share a large tent. Of course, a wooden wall would be constructed down the middle of the tent to preserve the proprieties.

The roof of the tent, under which the vertical boards of the wall would have to fit, curved down from its peak toward the tent walls. In the first year of the joint camp operation, the camp directors used six large rectangular pieces of plywood to form the wall. Unfortunately the flat tops of these rectangles did not lie flush with the curving roof of the tent, leaving ample space for enterprising campers of each camp to indulge their curiosity. On the positive side, applications shot up for the following summer. In the second year, the camp directors improved the wall by using a dozen planks to replace the plywood wall. These made a much better fit with the sloping roof

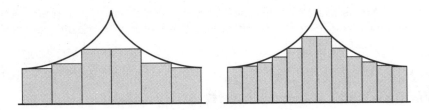

Figure 22.11 Two stages in the construction of the wall.

of the tent. However, the taller campers, standing on the top bunks, were still able to peek through the narrow gaps between the tops of the planks and the tent roof. The threat of lawsuits hung in the air. In the third year applications held steady, though most included requests for upper bunks. This time the camp directors rebuilt the wall with 1-inch strips of wood. These nearly filled the gap between their tops and the tent, so that even the most wily campers could see nothing, except an occasional eye on the other side. The crisis was resolved. Unfortunately, applications subsequently plummeted, and Camp Wantaugh-Peak was no more.

If the camp had not gone under and had continued year after year to refine its wall by constructing it with thinner and thinner strips of wood, then it would have been following the process used to construct a definite integral from Riemann sums (see Figure 22.11). In a Riemann sum, the area of the wooden planks used to build the wall approximates the area across the tent. As the planks are taken thinner and thinner, the wall fills up the entire area. The limiting area of all these planks is used to give the rigorous definition of the definite integral.

Technical Definition of the Definite Integral

$$\int_a^b f(x)\, dx \; = \; \lim_{n \to \infty} \left(\frac{b-a}{n} \right) [f(x_1) + f(x_2) + \cdots + f(x_n)]$$

Now notation rears its ugly head. Nobody ever writes it like this. Writing $f(x_1) + f(x_2) + \cdots + f(x_n)$ is okay for amateurs, but keep it up and there's a danger that everybody will understand this stuff. So, instead, the notation $\sum$ and Δx were introduced to keep out the uninitiated. The symbol $\sum$ (Greek letter sigma) is used for summation:

$$\sum_{i=1}^{n} f(x_i)$$

This means the same thing as $f(x_1) + f(x_2) + \cdots + f(x_n)$. The letter i is called the *index of summation*. The term Δx is used to denote the width of the thin rectangles we use in our approximation, $\Delta x = \dfrac{b-a}{n}$.

Our definition of the integral now looks *much* more impressive:

$$\int_a^b f(x)\,dx = \lim_{n\to\infty} \sum_{i=1}^{n} f(x_i)\,\Delta x$$

There are all kinds of variations on this. We can use any of the other approximation formulas to start off with. Or we can cut $[a,b]$ into pieces of unequal length. Or we can evaluate $f(x)$ in the middle of the subintervals instead of at the left endpoint. And lots of other stuff like that. As long as $f(x)$ is not a pathological function, we'll end up with the same number. Using the rectangle rule is the easiest way, and the most common. Here's the most complicated variation you're likely to see:

$$\int_a^b f(x)\,dx = \lim_{n\to\infty} \sum_{i=1}^{n} f(c_i)(x_i - x_{i-1})$$

where c_i is any point in the subinterval $[x_i, x_{i+1}]$.

What's the difference? In the complicated version, we can cut the interval $[a,b]$ into n *unequal* pieces, whose widths are given by $(x_i - x_{i-1})$. We can also evaluate f at any point c_i in the subinterval $[x_{i-1}, x_i]$. This works as long as the size of all our subintervals approaches zero as n goes to infinity and f is cool, meaning not too discontinuous.

And, of course, if we stop before we get to the limit on the right-hand side, we have a *Riemann sum*. So

$$\sum_{i=1}^{114} f(x_i)\,\Delta x$$

is a Riemann sum approximation of

$$\int_a^b f(x)\,dx$$

At the risk of sounding boring (hey, we're daredevils), we want to point out that your love and appreciation for the fundamental theorem should be growing exponentially here. The *definition* of the definite integral is this nasty creature involving limits, sums, and Δ's. Without the fundamental theorem, you'd be stuck having to use this formula to calculate these things with your bare hands. The fundamental theorem saves you from this hideous fate, allowing you to calculate integrals by just taking an antiderivative and evaluating it at the limits of integration.

WARNING: If a function is really nasty, it may not have an integral! But if the function is continuous or even continuous except for finitely many jumps, then all is hunky-dory, and we say the function is *integrable*.

Well, that was a lot of work. How about a break?

Fable of the Lecture for General Audiences. A not-to-be-named university has a lecture series aimed at a general audience and given by faculty. Many years have gone by, but no one from the math department has ever been asked to give one of the talks, for fear that a mathematician wouldn't be able to give a talk for an audience with no math background. But finally, it could be avoided no longer. So they ask this math professor to give a talk, but they say to him, "Look, this is for a very general audience. These people have not even had calculus, so they don't know about derivatives and integrals."

"No problem," says the professor, "I get the idea."

The time for the talk arrives, and the professor is introduced and stands up behind the podium to begin his talk. "Take the definite integral of a function $f(x)$ over the interval from a to b." The organizers, sitting in the front row, are horrified. The faces in the audience register confusion. But the professor looks down reassuringly at the organizers before continuing. "For those of you unfamiliar with the definite integral, just think of it as the limit of a Riemann sum...."

Modeling: From Toy Planes to the Runway

Math problems generated by real life (as opposed to ones cooked up for examples) are messy. Suppose you're given a problem, and told to make a mathematical model for it. You will need to

1. Clean it up. That is, throw out all the information that's irrelevant.

2. See if you can turn the remaining information into a math problem.

3. Solve the problem you came up with in step 2.

Each of these steps has its pitfalls, and you may have to make several tries before you can complete all three steps. The results are worth it: Good mathematical models are amazingly successful at predicting real life. This is true despite the fact that in order to get through steps 2 and 3 successfully, you often have to throw out tremendous amounts of apparently relevant information in step 1, and you often have to make vast simplifying assumptions before you can write down equations in step 2 which you have any hope of solving in step 3.

In doing your modeling, you will have an advantage over Newton and Leibniz and their gang. You will have access to computers and calculators.

This will enable you to quickly do calculations and computations that were out of reach even for great minds like theirs.

Modeling Tale   A mathematician went to the racetrack with her two pals, a physicist and a statistician. They decided they would each bet a hundred dollars on a horse in the big race and see who did best.

The statistician went to a newsstand and bought the past performance records for each of the horses. Whipping out her laptop computer, she programmed in all available variables and got a probability distribution for the winning chances of each horse. Comparing this with the posted odds, she bet the entire wad on the number 3 horse, Psychic Folly.

The physicist estimated the mass and size of each horse, its relative muscle ratio and the amount of work it could do (one horsepower, as it turns out), the friction coefficient of the track, and the wind force. He then bet all his cash on number 4, Cold Fusion.

The mathematician went to a quiet spot and stared off into space for a while, ignoring the track and the horses. She then bet all her money on the number 5 horse, Fermat's Last.

Sure enough, at the end of the race Fermat's Last had coasted to an easy win. "How did you do that?" demanded her two friends.

"Well," she explained. "If you came to my modeling lectures in calculus class you would have known what to do. First of all, I assumed that each horse was a perfect sphere..."

Moral: Even if the assumptions in a modeling problem seem to be far from reality, they may retain enough information to let you get some understanding of the behavior of the problem. Maybe enough to win the race.

23.1 Real-life problem

Problem   *It's 105 degrees in the shade, there's no shade, it's 11:30 in the morning, and you're sweating up on the roof of your house, holding a green balloon filled with tapioca, which you're planning to drop on your roommate's head as soon as he walks out the door. He's ruined your life, he deserves a little tapioca. The roof is 30 feet off the ground and the tapioca is beginning to cook inside the balloon. How fast will the pudding be moving when it hits your roommate's head? How much of a mess will it make?*

We want to ignore all the truly irrelevant stuff (we know the fact that he's ruined your life is the WHOLE REASON you're up on the roof, but it's not going to tell you how much of a mess this will make) and to simplify all the other information enough so that we can solve the problem.

What can we ignore? How about the temperature? Of course this is the reason the tapioca is beginning to cook, but we're going to ignore that, so we'll throw out the temperature information. The facts that it's 11:30, you're broiling up there, and you haven't had your first cup of coffee yet may seem important to you, but we're going to throw those out too.

Does it matter that the balloon is filled with tapioca, not water? Quite likely—tapioca isn't going to splash as far as water, and it's going to be a lot stickier to clean up. So we'll hold on to that piece of info. How about the shape of the balloon? That's trickier. It probably does matter what shape the balloon is, but we don't really have enough specific information about its shape to fix it. So we'll assume it's round. The fact that the roof is 30 feet high sounds important. But how tall are you? Are you going to be dropping the balloon from the roofline? Or from the highest point your arms can reach? For that matter, how tall is your roommate? We don't know any of this stuff, so we'll guess that you're going to be dropping it from, say, 6 feet above the roofline onto your roommate's head, which, handily, is 6 feet above the ground. In other words, you'll be dropping it 30 feet. Blam.

How fast will it be moving when it hits him? Now we can pull out some big guns: We'll use the fact that the acceleration due to gravity is $-32\,\text{ft/sec}^2$. This is a primo example of mathematical modeling at work. There should be WARNING WARNING written all over this fact. Would this be useful information if, instead of your loathsome roommate, your sweetest sweetie were exiting the house and you were planning to shower her with rose petals? No, rose petals and tapioca-filled balloons don't fall at the same speed, because rose petals get wafted around by the air like crazy; with tapioca, no wafts, or none to speak of. So we assume that the acceleration of the tapioca balloon is precisely $-32\,\text{ft/sec}^2$, and we won't be too far wrong. Write $a(t) = -32$.

The velocity (or speed) of the balloon is some antiderivative of the acceleration, so $v(t) = -32t + C$ for some C. What's C? Well, we know the velocity of the balloon is zero at that critical moment when you're holding it up in the air, pondering the meaning of life, the inevitable decline of civilization, and whether or not you're really going to drop that sucker on your roommate's head. So $v(0) = 0$. That means $C = 0$. So $v(t) = -32t$.

Things are looking up. If we can just figure out WHEN the balloon hits his head, we'll be able to tell how fast it'll be moving. It will hit his head after it's traveled 30 feet, right? So we need to know the distance function, which is an antiderivative of the velocity function: $d(t) = -16t^2 + D$. What's $d(0)$? $d(0)$ is 36 feet, since we assumed you're dropping it from 6 feet above the roofline. So $d(t) = -16t^2 + 36$. When will it hit his head? When $d(t) = 6$, or when $-16t^2 = -30$ or when $t^2 = {}^{15}\!/_{8}$ or when $t \approx 1.37$ seconds.

How fast will it be moving when it hits? Well, $v(1.37) = -43.8$ ft/sec. A satisfying speed. Likely to make a total mess. But that brings us to the next question. How much of a mess?

First interpret the question so that it makes sense. Remembering that we've assumed the balloon is round, imagine dropping it from 3 feet off the ground. When it hits, it bursts, scattering tapioca all over the place. Actually not quite all over the place—it's made a circle of tapioca, right? We'll call the radius of this circle the splash radius. Let's decide that the question: "How much of a mess will it make?" is equivalent to "What will the splash radius be when it hits?" Unfortunately we can't really answer the question about the splash radius of the roommate problem, because we don't have enough information. We don't know the splash radius of any tapioca balloon from any height. And this is not likely to be an experiment that you're going to be able to repeat.

We need some additional information. So we'll send your other roommate, the nice one, back to the lab (kitchen) for a quick test run. He drops a tapioca balloon from a height of 9 feet (he climbed onto the counter) onto the kitchen floor, making a truly unbelievable mess, and measures that it has a splash radius of 7 feet. Using the same kind of argument as before, we can calculate that the speed of the balloon when it hit the floor was −24 ft/sec.

Now we'll make a guess about the relation between the splash radius and the speed of the balloon on impact. We'll assume that THEY ARE PROPORTIONAL. That means that there is some mystery number, call it K, so that:

$$\text{splash radius} = K(\text{speed})$$

This is mathematical modeling again—in the real world, we might have to reconsider this hypothesis. But it seems reasonable on a first pass.

We have enough info from our test run to figure out K:

$$7 = K(-24) \quad \text{so} \quad K = -7/24$$

Presto: The splash radius is $-7/24$ times the impact velocity. The impact velocity on your roommate's head is going to be −43.8, so the splash radius will be $(-7/24)(-43.8) \approx 12$ feet. By the time he finishes cleaning it up you should be well on your way to Canada.

Exponents and Logarithms: A Review of All That "*e*" Hoopla

24.1 Exponents

Okay, this is probably all old hat, but let's go over it anyway, just to be on the safe side. Exponents are the little numbers that sit like parrots on the shoulders of numbers and variables. Like parrots they have their own set of rules. The rules for parrots are (1) make fun of your owners by mimicking their most hackneyed catch phrases and (2) mate with other parrots. The rules for exponents are not too different. Here's how they look:

$$2^2 \times 2^3 = 2^5 \qquad \text{and} \qquad (2^3)^5 = 2^{15}$$

These are easy to remember if we think of 2^3 as 2 multiplied by itself 3 times, and so on.

$$2^2 \times 2^3 = (2 \times 2)(2 \times 2 \times 2) = 2 \times 2 \times 2 \times 2 \times 2$$

We also need to know that

$$2^{-3} = \frac{1}{2 \times 2 \times 2}$$

We then get rules like

$$x^{a+b} = x^a \times x^b$$

$$(x^a)^b = x^{ab}$$

Nothing too fancy, but good to know. What about $x^a + x^b$? We might be tempted to write that in a simpler form, say as x^c, but don't even think about it. Can't be done neatly. For example, $2^2 + 2^1 = 6$, which is not a nice power of 2.

What else are exponents good for? Check this out: Another way of writing $\sqrt{x}$ is $x^{1/2}$. Why? Well,

$$x^{1/2} x^{1/2} = x^{1/2+1/2} = x^1 = x = \sqrt{x}\,\sqrt{x}$$

So $x^{1/2}$ and $\sqrt{x}$ are the same.

All these rules work just the same for any exponents, whether they are whole numbers or not.

Also note that any number to the power 0 gives 1, as in $(345{,}762)^0 = 1$. There is one exception: 0^0 is such a zero that it's not even a number.

The number 2 in 2^3 is called the "base." We could just as well look at powers of the base 10 such as 10^3 or $10^{1/2}$. We could even look at powers of fractions and irrational numbers, including the strange number called "e" whose value is about 2.7182. This turns out to be a good idea, odd as it may seem.

Good Question: Why do you use the letter e for 2.718...?
Wrong answer: a through d were already taken.
Right answer: It's the first letter in exponential.

An English mathematician memorized the first 10,000 digits of e. He and his wife would practice by saying 100-digit sequences back and forth to each other as if they were having a numerical conversation. Why did they do it? Maybe there was nothing good on English TV.

Group Project: Get together with two of your classmates and memorize the first 100,000 digits of e. If you don't have time, try it for the first 50,000. If you're really busy, use the number 3 instead of e.

Figure 24.1 shows the graph of e^x. What are its important features?

1. It passes through $(0, 1)$, since $e^0 = 1$.

2. It is always positive, that is, $e^x > 0$ for all x.

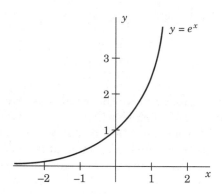

Figure 24.1 The graph of e^x.

3. It is always increasing.

4. It increases *very* fast. That's why really fast growth is called "exponential growth".

24.2 Logarithms

Taking a logarithm is the opposite of raising to a power. So if you say $2^3 = 8$, then in this opposite world they say $\log_2 8 = 3$. Here, 2 is again called the base.

How to remember this? Take the base across, like this:

$$2^3 = 8 \Rightarrow 3 = \log_2 8$$

If the base is the number e, then we write the log function as $\ln x$, and call it the natural log.

$$\ln x = \log_e x$$

Usually we work with $\ln x$ in calculus, rather than with a log to a base other than e, but the basic rules are the same with any base.

Now of course everyone has to have rules in their lives, and logarithms are no exception. But unlike a lot of other mathematical functions, logarithms dance to the beat of a different drum, one that beats a log rhythm. For instance:

$$\ln (a\,b) = \ln a + \ln b$$

$$\ln (a^b) = b \ln a$$

$$\ln \left(\frac{1}{a} \right) = -\ln a$$

You should get to know and love these rules. As we will see, they will save us tremendous amounts of work shortly.

Exponentials and logarithms are inverses. That means that one undoes what the other one does, much like a dog and a pooper-scooper. At the end of the walk, no one can see where the dog came and went, so to speak.

$$2^{\log_2 x} = x$$

and

$$\log_2 (2^x) = x$$

This works just as well if you replace 2 with b:

$$b^{\log_b x} = x$$

and

$$\log_b (b^x) = x$$

for any $b > 0$.

In particular, if you use e as the base and remember that $\log_e x$ is written as $\ln x$, then

$$\boxed{e^{\ln x} = x}$$

and

$$\boxed{\ln (e^x) = x}$$

When $x = 1$ or $x = 0$, we see two key facts:

$$\boxed{\ln e = 1 \quad \text{and} \quad \ln 1 = 0}$$

Let's take a look at the graph of $y = \ln x$ (Figure 24.2). Important features:

1. $(1, 0)$ and $(e, 1)$ are on the graph.

2. $\displaystyle\lim_{x \to +\infty} \ln x = \infty$.

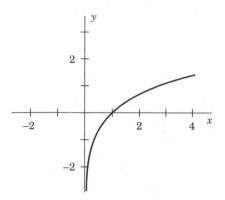

Figure 24.2 Graph of $y = \ln x$.

3. $\lim\limits_{x \to 0^+} \ln x = -\infty$.

4. $\ln x$ IS ONLY DEFINED FOR $x > 0$.

5. $\ln x$ is always increasing.

6. Since $\ln x$ is the inverse of e^x, its graph is the reflection of the graph of e^x over the diagonal line $y = x$ (see Figure 24.3).

If you want, you can just remember one of the two graphs and reflect it over the line $y = x$ to get the other. But make sure you can figure out which is which.

A variety of problems can be assigned on manipulating exponentials and logs. Even though they do not involve derivatives or integrals, they are considered fair game. Here's a sample.

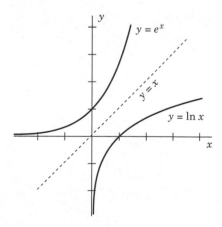

Figure 24.3 $y = e^x$ and $y = \ln x$ are inverses of one another.

Problem   *Show that*

$$\frac{\ln 3}{\ln 2} = \log_2 3$$

Well, let's set $x = \dfrac{\ln 3}{\ln 2}$. Then

$$x \ln 2 = \ln 3$$

Using one of the laws for logarithms,

$$\ln (2^x) = \ln 3$$

Exponentiate both sides

$$e^{\ln(2^x)} = e^{\ln 3}$$

and, since exponentiation is the inverse of ln and undoes it,

$$2^x = 3$$

So, taking $\log_2$ of both sides,

$$x = \log_2 3$$

just what we wanted to show.

Doing That Calc Thing to Exponents and Logs

Now let's take a look at the calculus of the functions e^x and $\ln x$.

25.1 Differentiating e^x and its friends

Remember:

1. e is a number whose value is about 2.72.

2. e is used as the base for the exponential and logarithm functions, e^x and $\ln x$.

The function e^x has the following amazing property:

$$\frac{d}{dx}(e^x) = e^x$$

The derivative of e^x is just e^x again. That's kind of like giving birth to yourself, a very difficult event to achieve and one that would get you major tabloid coverage.

NEWS FLASH: "I AM MY OWN MOTHER," FUNCTION CLAIMS!

In fact, e^x and its multiples are the only functions that are their own derivatives. For this reason we put up with the use of the very strange number $2.718281828459\ldots$ as the base.

Example   *Find* $\dfrac{d}{dx}(e^{\sin x})$.

Notice that $e^{\sin x}$ is the composition of e^x with $\sin x$. So we'll use the chain rule to differentiate it.

$$\frac{d}{dx}e^{\sin x} = e^{\sin x}\frac{d}{dx}(\sin x) = e^{\sin x}\cos x.$$

The $e^{\sin x}$ is the derivative of the outside function e^u evaluated at the inside function $u = \sin x$ and the $\cos x$ is the derivative of the inside function.

25.2 Integrating e^x and its friends

You're going to like this part. Since $\dfrac{d}{dx}(e^x) = e^x$, we know immediately that

$$\boxed{\int e^x\,dx = e^x + C}$$

In other words, if e^x is its own mother, then it's its own child, too.

Now, we can find lots of other integrals using substitution. For instance,

$$\int e^{3x}\,dx = \frac{e^{3x}}{3} + C$$

(You know, $u = 3x$, $du = 3\,dx$, etc. Usual drill.)

More generally, it's worth remembering that

$$\int e^{kx}\,dx = \frac{e^{kx}}{k} + C \qquad \text{for any constant } k \text{ other than 0.}$$

We could also use substitution or eyeball to see that

$$\int e^{x^4}x^3\,dx = \frac{e^{x^4}}{4} + C$$

25.3 Differentiating the natural log

Let's not beat around the bush. Here you go:

$$\frac{d}{dx}(\ln x) = \frac{1}{x}$$

Pretty amazing that the derivative of an ugly function like $\ln x$ would be a pretty function like $\frac{1}{x}$. It's kind of a Cinderella story for functions, but without the pumpkins, glass slipper, and raging royal hormones.

Now, suppose we want to differentiate $\ln(\sqrt{x})$. We have two options:

Option 1:   Just do it by the chain rule.

$$\frac{d}{dx}\ln(\sqrt{x}) = \frac{1}{\sqrt{x}}\frac{1}{2\sqrt{x}} = \frac{1}{2x}$$

Option 2:   Use the fact that

$$\ln(\sqrt{x}) = \ln(x^{1/2}) = \frac{1}{2}\ln x$$

Then,

$$\frac{d}{dx}\ln(\sqrt{x}) = \frac{d}{dx}\left(\frac{1}{2}\ln x\right) = \frac{1}{2x}$$

In general, the chain rule says

$$\boxed{\frac{d}{dx}(\ln g(x)) = \frac{g'(x)}{g(x)}}$$

So for instance,

$$\frac{d}{dx}(\ln(x^3 - 7)) = \frac{3x^2}{x^3 - 7}$$

25.4 Working with other bases

You can choose almost any number as the base for logarithms and exponentials. Since we have 10 fingers we often think of 10^x and the corresponding logarithm function $\log_{10} x$. In the old days, schoolchildren got lots of practice using tables of logarithms with base 10. If we had two fingers we might have used base 2. Come to think of it, we do have two arms. Anyway, these days we only use one finger to peck at the keys of a calculator and the function $\log_{10} x$ is disappearing from the world.

Let's look at the derivatives of b^x and $\log_b x$. In particular, let's start with 2^x. Now you might be tempted to say

$$\frac{d}{dx}(2^x) = x\, 2^{x-1}$$

BUT THAT IS WRONG. You have to learn to control your baser instincts. The power rule doesn't apply when the base is a constant and the exponent is a variable. Instead, the correct answer is

$$\boxed{\frac{d}{dx}(2^x) = 2^x \ln 2}$$

A factor of $\ln 2$ multiplies the derivative. Why $\ln 2$, for goodness sake, you ask? Well, look. $2 = e^{\ln 2}$, so $2^x = (e^{\ln 2})^x = e^{x(\ln 2)} = e^{\ln(2^x)}$ by the rules for logarithms. Applying the chain rule, we have

$$\frac{d}{dx} 2^x = \frac{d}{dx} e^{\ln(2^x)} = e^{\ln(2^x)} \frac{d}{dx}(\ln(2^x)) = e^{\ln(2^x)} \frac{d}{dx}(x \ln 2)$$
$$= e^{\ln(2^x)} \ln 2 = 2^x \ln 2$$

The same argument gives the general case:

$$\boxed{\frac{d}{dx}(a^x) = a^x \ln a}$$

In the next chapter, we will introduce a second method for finding $\frac{d}{dx}(a^x)$.

Now, what about differentiating $\log_b x$? We want to find $\frac{dy}{dx}$ where $y = \log_b x$. But, $y = \log_b x$ implies $b^y = x$. Let's implicitly differentiate this equation. We get

$$b^y \ln b \frac{dy}{dx} = 1$$

$$\frac{dy}{dx} = \frac{1}{b^y \ln b} = \frac{1}{x \ln b}$$

We've just shown that

$$\frac{d}{dx}(\log_b x) = \frac{1}{x \ln b}$$

An extra factor of $\ln b$ occurs for exponentials with base b. Notice that since $\ln e = 1$, there is no extra factor exactly when the base is e. Makes you really appreciate e, doesn't it?

25.5 Integrals and the natural log

We know, you're expecting us to begin this section by telling you what $\int \ln x \, dx$ equals. But that's not really what's important here. If you really must know, it equals $x \ln x - x + C$, but few instructors expect you to remember that (although it can be worked out using integration by parts, a section coming up shortly, and there, it makes a good problem.)

No, in this section we would just like to reverse the formula

$$\frac{d}{dx}(\ln x) = \frac{1}{x}$$

to obtain one of the most famous formulas in all calculus:

$$\int \frac{1}{x} \, dx = \ln |x| + C$$

There are a couple of things to notice here. First, by integrating a function $1/x$ that appears to have nothing whatsoever to do with e, wouldn't recognize e if e bit it on the nose, we get the log base e. Just another way of saying, "e, you're mighty special."

Second, you will notice that somehow, the x in the solution picked up absolute value signs. That's because the natural log function is only defined for positive values of x, so if x were negative and we didn't have the absolute value signs, it would make no sense. Should you worry about the absolute value signs? If you're going for the A+, *yes,* otherwise forget about them.

Problem   *Find* $\int \tan x \, dx$.

Now $\tan x$ is just $\sin x / \cos x$.

$$\int \tan x \, dx = \int \frac{\sin x}{\cos x} \, dx$$

This looks ripe for a u substitution. Let's take $u = \cos x$, so $du = -\sin x \, dx$ and

$$\int \frac{\sin x}{\cos x} \, dx = \int \frac{-1}{u} \, du$$

$$= -\ln |u| + C$$

$$= -\ln |\cos x| + C$$

$$= \ln \frac{1}{|\cos x|} + C$$

$$= \ln |\sec x| + C$$

Now, was that as good for you as it was for us?

Logarithmic Differentiation: Making the Hard Stuff Easy

Good old logarithmic differentiation. (If you are really cool, you call it "log diff.") All it means is that if you have a function you want to differentiate, you take the natural log of both sides of an equation before you take the derivative.

Why would you want to do that? Let's look at a couple of examples.

Say we want to find the derivative of

$$f(x) = (x^2 - 3)(x^3 - 4)(x^7 - 5)(x^2 - 6)$$

Well, we could just apply the product rule a whole bunch of times, and blow our afternoon (okay so we are exaggerating a little; blow the next 15 minutes at least) and probably make a mistake. If we want to save ourselves a whole lot of agony, we could first take logs of both sides.

$$\ln f(x) = \ln[(x^2 - 3)(x^3 - 4)(x^7 - 5)(x^2 - 6)]$$

But then, by the properties of ln, we have

$$\ln f(x) = \ln(x^2 - 3) + \ln(x^3 - 4) + \ln(x^7 - 5) + \ln(x^2 - 6)$$

Our nasty product has become a pretty sum. This is just what logs were invented for.

Now we'll differentiate both sides, using the fact that the chain rule says the derivative of the left side is $f'(x)/f(x)$.

$$\frac{f'(x)}{f(x)} = \frac{2x}{x^2 - 3} + \frac{3x^2}{x^3 - 4} + \frac{7x^6}{x^7 - 5} + \frac{2x}{x^2 - 6}$$

That was pretty easy, but unfortunately, it isn't the answer that we wanted. Our goal was to find $f'(x)$. Not to worry. If we multiply both sides of the equation by $f(x)$, which we know from the statement of the problem, we will have $f'(x)$ by itself.

$$f'(x) = f(x)\left[\frac{2x}{x^2 - 3} + \frac{3x^2}{x^3 - 4} + \frac{7x^6}{x^7 - 5} + \frac{2x}{x^2 - 6}\right]$$

$$= (x^2 - 3)(x^3 - 4)(x^7 - 5)(x^2 - 6)\left(\frac{2x}{x^2 - 3} + \frac{3x^2}{x^3 - 4} + \frac{7x^6}{x^7 - 5} + \frac{2x}{x^2 - 6}\right)$$

DONE, FINITO, AND WITH A LOT LESS TIME AND CONFUSION THAN WOULD HAVE RESULTED FROM TRYING TO USE THE PRODUCT RULE OR, WORSE YET, MULTIPLYING THE THING OUT.

By the way, don't bother multiplying the answer out, or you'll waste some of the time we just saved.

Let's review the game plan. We have a function $f(x)$ that is a messy product. We want to know the derivative.

1. Take ln of both sides.

2. Simplify the right side using laws of logs.

3. Take derivatives of both sides. We always get $f'(x)/f(x)$ on the left side.

4. Multiply through by $f(x)$, which you know from the beginning of the problem, to get $f'(x)$ all by itself.

Common Mistake   Many people forget the last step and do not multiply through by $f(x)$ at the end.

What else is log diff good for? Check out this alternative easy way to differentiate 2^x.

Let $f(x) = 2^x$. We want to find $f'(x)$.
First take logs of both sides:

$$\ln(f(x)) = \ln(2^x)$$

By the laws of logs,

$$\ln(f(x)) = x \ln 2$$

Taking the derivative of both sides and using the fact that $\ln 2$ is just a constant, we have

$$\frac{f'(x)}{f(x)} = \ln 2$$

So

$$f'(x) = f(x)\ln 2 = 2^x \ln 2$$

What is particularly astonishing about this fact is that we started with a function that had no logs in it, but when we differentiated it, suddenly an "ln" appeared out of thin air. This is why we call it the natural log. It appears in a lot of contexts au naturel.

Exponential Growth and Decay: Rise and Fall of Slime

Exponential growth problems often tend to the disgusting, involving either the growth of bacteria cultures or nuclear holocaust. But even the distasteful side of life can be explained by mathematics.

Basically, any situation where the rate of change of a function is proportional to the value of the function is an example of exponential growth or decay. That's a mouthful. But for example, population growth fits this pattern. The rate of growth of a population of rabbits is greater when you have more rabbits. Radioactive decay behaves in a similar manner, though the amount of material that you have is shrinking over time instead of growing. The less radioactive material you have, the slower it decays each minute. Letting N be the number of rabbits or the amount of plutonium that we have at any given time, the fact that the rate of change of the function (what we normally call the derivative of the function with respect to time) is proportional to the value of the function can be stated mathematically as

$$\frac{dN}{dt} = kN$$

Here, k is just the proportionality constant, with actual value depending on the problem. In the case of a growth problem, such as the rabbit population,

k will be positive. This is because dN/dt must be positive since the function giving the number of rabbits in the population N is an increasing function.

In the case of a decay problem, such as radioactive decay, k will be negative. This is because dN/dt must be negative, so that the function giving the amount of radioactive material at any given time N is a decreasing function.

This equation

$$\frac{dN}{dt} = kN$$

is a so-called *differential equation*. It's called a differential equation because it is an equation that involves differentiation. It is just about the simplest differential equation you can come up with. (All right, all right, there are one or two simpler ones. Take $dN/dt = 0$, for instance.)

Unlike most of the differential equations out there, this one is solvable. That is to say, we can figure out exactly the general form of a function N that satisfies this equation.

First we "separate the variables," putting everything that involves N on the left side of the equation and everything that involves t on the right side of the equation. We even treat the derivative dN/dt as if it were a fraction, splitting the dN from the dt:

$$\frac{dN}{dt} = kN$$

$$\frac{dN}{N} = k\,dt$$

Now we integrate both sides:

$$\int \frac{dN}{N} = \int k\,dt$$

$$\ln N = kt + C$$

(Notice that we didn't bother with the absolute values on the N, since we know N is positive here.)

Applying the exponential function to both sides, we obtain:

$$N = e^{kt+C} = e^{kt}e^C = Ae^{kt}$$

where we replace e^C by A, since e^C is just some constant anyway. Why use a fancy name like e^C for it, when A works just as well?

So $N = Ae^{kt}$. Notice that at time $t = 0$, $N(0) = Ae^{k0} = A$. Since $N(0)$ is the initial amount that we started with, A is just this initial amount, usually written as N_0.

That means that our general formula for the solution to the differential equation

$$\frac{dN}{dt} = kN$$

is given by

$$\boxed{N(t) = N_0e^{kt}}$$

This is called our *exponential growth* or *exponential decay* equation. Let's apply it in some representative examples.

Sanitation Problem *Suppose a colony of bacteria is growing in the corner of your shower stall. On June 1, there are 1 million bacteria there. By July 1, there are 7.5 million. Your shower stall can contain a total of 1 billion bacteria. When will you have to start taking showers down at the gym?*

Okay, we know what you're thinking. You're thinking, "Well actually, even before the entire shower stall is filled with bacteria, I'm not going to take showers in there. I mean, when it's half full, I'm not getting in there. And since I don't know exactly when I will be unwilling to get in there anymore, this problem is poorly stated and therefore I refuse to try to solve it."

But you have to be tough and get into the shower until the very end. THIS IS MATH. It's not for the lily-livered.

Solution We know that the number of bacteria in this colony is given by $N(t) = N_0e^{kt}$. Let's take June 1 as the time $t = 0$. Then, the initial number of bacteria is $N_0 = 1,000,000$. So we have

$$N(t) = 1,000,000e^{kt}$$

We now need to determine k. But we know that on July 1, which corresponds to $t = 30$, the population $N(30) = 7,500,000$. Plugging this into our function, and solving for k, we have

$$7,500,000 = N(30) = 1,000,000e^{k(30)}$$

$$7.5 = e^{k(30)}$$

$$\ln(7.5) = 30k$$

$$k = \frac{\ln(7.5)}{30} \approx 0.0672$$

Plugging that value for k into our original function, we now have

$$N(t) \approx 1,000,000e^{0.0672t}$$

This tells us the number of bacteria at any time. When will this equal 1,000,000,000? Set it equal to 1,000,000,000 and solve for t:

$$1,000,000,000 \approx 1,000,000e^{0.0672t}$$

$$1,000 \approx e^{0.0672t}$$

$$\ln(1,000) \approx 0.0672t$$

$$t \approx \frac{\ln(1,000)}{0.0672} \approx 102.8\,\text{days}$$

So, on September 10, that shower is one giant black slimy mass. Get ready to move.

Finding an Apartment in New York Problem   *A nuclear bomb set off by terrorists has made New York City uninhabitable. Many people argue that this is not a change. But suppose the residual cobalt levels are 100 times safe levels. If cobalt has a half-life of 5.37 years, how soon before people can live in New York again?*

Solution   This is called a decay problem. Not because our urban centers are in decay, but because radioactive material decays over time, so the remaining amount decreases. What do we do here? Let $N(t)$ be the amount of cobalt in the city at time t, t years after the initial explosion. Then $N(t) = N_0e^{kt}$, where N_0 is the initial amount of cobalt released into the city and k is our decay constant, which must be negative. In most decay-type problems, N_0 is given to you. But here, we don't even know that. However, we have been given the half-life of cobalt. We can use that to determine the decay constant k.

Since the half-life of cobalt is 5.37 years, we know that if we start with an initial amount of cobalt N_0, we will have $N_0/2$ of it left at the end of 5.37 years. Therefore $N_0/2 = N(5.37) = N_0e^{k\,5.37}$. So, solving for k:

$$\tfrac{1}{2} = e^{k\,5.37}$$

$$\ln(\tfrac{1}{2}) = k\,5.37$$

$$k = \frac{\ln(\tfrac{1}{2})}{5.37} \approx -0.129$$

Therefore, $N(t) = N_0 e^{-0.129t}$ gives the amount of remaining cobalt at any given time.

Now we want to know when it will be safe for people to return to New York. Initially, cobalt levels were 100 times safe levels, so we would like to know when the amount of cobalt will be $N_0/100$. At that time, the amount of cobalt will have dropped down to a safe level. How do we determine when that time is? We just set our function giving the level at any time equal to that amount and solve for t:

$$N(t) = N_0 e^{-0.129t} = \frac{N_0}{100}$$

$$e^{-0.129t} = \tfrac{1}{100}$$

$$-0.129t = \ln(\tfrac{1}{100})$$

$$t = \frac{1}{-0.129}\ln\left(\frac{1}{100}\right) \approx 35.7\,\text{years}$$

Not bad, although personally we're waiting at least 36 years, just to be on the safe side.

Case of the Big Cheese   "Well, look who the cat dragged in," said Sergeant Woffle, as Detective Horns walked into the room. "If it isn't Sheer Luck Horns."

"Still talking in clichés, I see," said Horns. "And the sarcasm needs work. What's the m.o.?" He looked down at the crumpled body lying in a heap on the floor. Woffle shrugged.

"He's Hiram Fentley, heir to the Fentley feta cheese fortune. Found dead at 2:30 A.M. Somebody decided they had had enough Fentley."

"What else?"

"Body temperature at 3:00 A.M. was 85° and at 4:00 A.M. was 78°. Don't ask me how I got it. You don't want to know."

"All right, Sarge, so when did he die?"

"Don't ask me, Horns, you're the detective."

"Well then, let me give you a free lesson, and maybe you won't spend the rest of your career taking temperatures. If a body is cooling in a room with air temperature R, then the rate of change of the temperature of the body, call it 'T,' is proportional to the difference between the body temperature and the room temperature, namely, $T - R$.

"So the differential equation governing temperature T says:

$$\frac{dT}{dt} = k(T - R)$$

where R is the room temperature. Of course, this isn't the standard differential equation that governs exponential growth or decay."

"Of course."

"But as is done with the standard equation, we can separate the variables, putting all T's on the left, and all t's on the right:

$$\frac{dT}{T - R} = k \, dt$$

Integrating both sides, we get

$$\int \frac{dT}{T - R} = \int k \, dt$$

$$\ln(T - R) = kt + C$$

$$T - R = e^{kt+C} = e^{kt}e^{C} = e^{kt}A$$

where we've replaced e^{C} with A just because it is prettier."

"You always had an eye for the capital letters," interjected Woffle. Horns ignored him.

"So

$$T(t) = Ae^{kt} + R$$

where R is the room temperature, a constant. In our case, we have $T(t) = Ae^{kt} + 70$."

"I follow you."

"Sure you do. Now, we'll use the particular temperature readings that you took to determine both of the constants A and k. It's our choice as to when $t = 0$, so let's choose $t = 0$ when the first temperature reading was taken, at 3:00 A.M.

"Then we know $85 = T(0) = Ae^{(k)0} + 70$. So $85 = A + 70$ and $A = 15$. Therefore,

$$T(t) = 15e^{kt} + 70$$

"We also know $78 = T(1) = 15e^k + 70$. So $15e^k = 8$.

$$e^k = 8/15$$

$$k = \ln(8/15) \approx -0.6286$$

as can be seen by these logarithmic tables I always carry with me. So our function governing the body temperature is:

$$T(t) = 15e^{-0.6286t} + 70$$

"Assuming he didn't just step out of a sauna, Fentley's temperature was 98.6° at the instant he was murdered. So we set our temperature function equal to this and solve for t to determine exactly when he was murdered.

$$98.6 = 15e^{-0.6286t} + 70$$

$$28.6 = 15e^{-0.6286t}$$

$$\frac{28.6}{15} = e^{-0.6286t}$$

$$\ln\left(\frac{28.6}{15}\right) = -0.6286t$$

So

$$t = \frac{\ln(28.6/15)}{-0.6286} \approx -1.02 \text{ hours."}$$

"So what does all that mean?" said Woffle.

"It means," said Horns, "that Hiram bit the big one around 2:00 A.M."

"Okay, Horns, that's good, but how did he die?"

Horns leaned over, wedged his fingers in the dead man's mouth and yanked out a large piece of cheese. "If I don't miss my guess," he said, "Hiram suffocated to death on this."

"You mean he just choked on a piece of cheese he was eating?"

"No, Woffle, that's not what I mean. No person in his right mind, let alone a cheese baron, would put a hunk of cheese this big in his mouth."

"So, someone stuffed it in there? But it's too big. How could it get in?"
"Ah, but don't you see, Woffle, there is one way to get it in."
"What's that?"
"It was a simple twist of feta."

Well, after all this violence, slime, and nuclear holocaust, let's see if we can come up with a nice sweet growth or decay problem, without this dark side.

Gardening Shop Problem   *A lovely elderly woman named Adelaide has invested her life's savings in a gardening shop specializing in daisies. Unfortunately, due to a drought, the daisies do badly, and the shop is going to go under unless it has an infusion of cash. Adelaide has a choice of borrowing $5000 from a loan shark at 23% for one year, compounded continuously, or $5000 from her son at 24% for one year compounded quarterly. Assuming that her legs will be broken, in either case, if she doesn't pay up at the end of the year, which deal should she go with?*

Look, we're sorry about the broken legs part, but we just want to get across this idea that growth and decay problems are about the slimy underbelly of life, the evil that lurks just beneath the surface of our bucolic everyday existence. Anyway, if we get her the better deal, maybe she'll be able to pay off the loan.

Solution   First, let's see what her son is charging her for one year. He wants 24% for the year, which is 6% per quarter. At the end of the first quarter, she will owe her son $5000(1 + 0.06)$. Since we are compounding quarterly, it is on this amount that we compute the interest for the next quarter. At the end of the second quarter, she will owe $5000(1 + 0.06)^2$. The amount she owes at the end of the third quarter will be 1.06 times this amount,

$$5000(1 + 0.06)^3$$

And at the end of the fourth quarter, which is the end of the year, she will owe

$$5000(1 + .06)^4 = \$6312.39$$

rounded up to the nearest penny. (Her son always rounds up.) While we are at it, let's just mention that if you are investing P dollars at a rate of interest of r (r expressed as a decimal) for a period of t years, compounded n times a year, then the amount of money that results is $P(1 + r/n)^{nt}$. The argument is essentially the same as the one we used above.

Now let's see what the loan shark is going to charge. Since the loan shark uses continuous compounding, the amount owed is

$$5000e^{0.23(1)} \approx \$6293.00$$

So she gets the better deal from the loan shark. It's a good thing she knows calculus. (Just so you know, it all worked out in the end; after she created a catchy homepage on the Web, her flower shop became a major corporation, gobbling up the family-run flower businesses all over the country, and eventually her son came to work for her where she humiliated him on a daily basis.)

General Formula to Remember   If P dollars is invested at an annual rate of interest r for a period of t years, compounded continuously, then the amount of money that results is

$$Pe^{rt}$$

Fancy-Pants Techniques of Integration

Now here is what we haven't yet admitted about integrals. Most integrals are wild and dirty creatures that haunt the darkest recesses of calculus, and resist all attempts at domestication. Uncivilized and not even housebroken, they shrink from the light of day and purposely wear the same underwear all week. It's not pretty.

As you can then imagine, it is a rare integral indeed that willingly allows itself to be integrated. Most fight and claw and threaten lawsuits. If we want to successfully trap and tame the wild integrals and teach them proper table manners, we need a variety of snares. That's what this chapter is all about, techniques of integration that greatly extend the set of integrals we can integrate.

Before we begin, keep in mind one caveat. As we cover each technique of integration in this chapter, all the problems in that section have to do with that method. However, when you get to the exam, there will be a list of integrals and no hints as to which technique to apply. So you want to be able to look an integral over and say, "Aha, clearly a case for partial fractions, Watson." And that's even if there is no one named Watson in the room. Being able to recognize which technique to apply is at least as important as being able to apply it.

Importance of Learning Techniques of Integration Two days after the exam, a student receives a message to see the professor. When he arrives, the professor says, "I have some bad news and some good news. The bad news is that you don't understand techniques of integration at all. You used integration by parts when you should have used substitution, you used trigonometric substitution when you should have used partial fractions. In sum, you completely blew the exam. Your score was so low that there's no way for you to pass the course. Having looked over your previous academic record, I'm sorry to have to tell you this will mean you will shortly be expelled from our institution of higher learning."

The student is aghast. "I can't believe it," he says. "This is just awful." Still stunned, he asks, "What's the good news?"

The professor leans over conspiratorially. "The Knicks are playing the Celtics on Friday night," he says, "and I got a courtside seat."

28.1 Integration method: Integration by parts

If $u(x)$ and $v(x)$ are functions of x, the product rule works great to find the derivative of their product.

Product Rule $(u(x)v(x))' = u'(x)v(x) + u(x)v'(x)$

What happens when we try to reverse it for a rule of integration? We get a rule called **integration by parts:**

$$\int u(x)v'(x)\,dx = u(x)v(x) - \int u'(x)v(x)\,dx$$

or, more compactly,

$$\boxed{\int uv'\,dx = uv - \int u'v\,dx}$$

By using the shortened forms

$$du = \frac{du}{dx}dx = u'\,dx$$

and

$$dv = \frac{dv}{dx}dx = v'\,dx$$

we get the usual form for integration by parts:

$$\int u\ dv\ =\ uv\ -\ \int v\ du$$

The trick in applying this powerful method is to decide which part of the original integral should be u and which part should be dv. This is akin to deciding where the part should go in your hair. What goes to the left of the part, and what goes to the right? Many an aspiring beautician has found that an inability to choose parts results in a lump of matted hair, stopping up the drain of their ambition. On the other hand, poor choices of parts when integrating could mean that a career in hairdressing looks pretty good. So let's focus on some examples.

Example *Find* $\int x \ln x\ dx.$

Let's try $u\ =\ \ln x$ and $dv\ =\ x\ dx$. What does integration by parts give? Make a chart:

$$u\ =\ \ln x \qquad dv\ =\ x\ dx$$
$$du\ =\ \frac{1}{x}\ dx \qquad v\ =\ \frac{x^2}{2}$$

We get v by integrating dv. In this case $\int x\ dx\ =\ x^2/2$, so $v\ =\ x^2/2$. We can choose the $+C$ term to be 0 since we will still have a $+C$ on the right side of the equation hidden inside the still undone integral $\int v\ du$.

$$\int x \ln x\ dx\ =\ \int u\ dv\ =\ uv\ -\ \int v\ du$$

$$=\ \frac{x^2}{2} \ln (x)\ -\ \int \frac{x^2}{2} \left(\frac{1}{x}\right) dx$$

$$=\ \frac{x^2}{2} \ln (x)\ -\ \int \frac{x}{2}\ dx$$

$$=\ \frac{x^2}{2} \ln (x)\ -\ \frac{x^2}{4}\ +\ C$$

And that's our answer. We can easily check that it's right by taking the derivative of our answer to see that we get $x \ln x$. Go ahead, take the derivative. We'll wait for you here.

Example   *What about* $\int \ln x \ dx?$

This is certainly a function we would like to be able to integrate, but where are u and dv in this one? They're hidden, lurking like roaches under the sink. We will pick $u = \ln x$ and $dv = dx$. Why this choice? Well, we all know how to differentiate $\ln x$. So it should be the u. Since there is nothing left other than the dx, that's all there is to go in the dv.

What does integration by parts give? Make yourself a chart:

$$u = \ln x \qquad dv = dx$$

$$du = \frac{1}{x} dx \qquad v = x$$

Using the integration by parts formula, we get

$$\int \ln x \ dx = x \ln x - \int 1 \ dx$$

$$= x \ln x - x + C$$

Integration by parts doesn't work for all functions, but when it does you get a great adrenaline rush.

How to Decide What Should Be u and What Should Be dv?

1. The dv has to be something you can integrate, which cuts the possibilities down substantially. The u is what's left over.

2. The resulting integral $\int v \ du$ that appears on the right after integration by parts should be easier to do than the integral you started with. If it's worse, then you might want to try a different u and dv.

28.2 Integration method: Trigonometric substitution

You may have heard that sine, cosine, and the other trigonometric functions satisfy certain secret identities. This is the source of the term "identity crisis," meaning an inability to remember a crucial identity on an exam. The identity you are least likely to forget is the famous

$$\sin^2 \theta + \cos^2 \theta = 1$$

There are many others, but this is the only one most people ever remember. We can reconstruct most of the others from it.

Trick (*An easy way to remember trigonometric identities*) Work from

$$\sin^2 \theta + \cos^2 \theta = 1$$

For example, by dividing every term in this equation by $\sin^2 \theta$,

$$\frac{\sin^2 \theta}{\sin^2 \theta} + \frac{\cos^2 \theta}{\sin^2 \theta} = \frac{1}{\sin^2 \theta}$$

and simplifying, we get

$$1 + \cot^2 \theta = \csc^2 \theta$$

We can also get

$$\tan^2 \theta + 1 = \sec^2 \theta$$

by dividing the first equation through by $\cos^2 \theta$ instead of $\sin^2 \theta$.

What are these identities good for? We can sometimes use them to find integrals of functions involving square roots or x^2's. The first example here is straightforward, and if you really understand the idea behind it, you will be able to do harder ones without getting confused.

Example *Calculate*

$$\int \frac{1}{\sqrt{1 - x^2}}\, dx$$

Looks bad, doesn't it? Both straight substitution and integration by parts produce a horrible mess with this one. So instead we replace x with something that will make the nasty $\sqrt{1 - x^2}$ disappear. Let

$$x = \sin \theta$$

Then

$$\sqrt{1 - x^2} = \sqrt{1 - \sin^2 \theta}$$

$$= \sqrt{\cos^2 \theta}$$

$$= \cos \theta$$

That'll certainly clean up the denominator. But just as in substitution, you can't let the dx hang out on its own, so we do the same kind of thing we did there:

$$\frac{dx}{d\theta} = \cos\theta$$

so

$$dx = \cos\theta \, d\theta$$

and the integral becomes

$$\int \frac{1}{\sqrt{1-x^2}} \, dx = \int \frac{1}{\cos\theta} \cos\theta \, d\theta$$

$$= \int 1 \, d\theta$$

The function we're integrating here is the constant function 1:

$$\int 1 \, d\theta = \theta + C$$

Now this is *not* the answer—we have to go back to the beginning to change that θ back into x's. Since

$$x = \sin\theta$$

we can solve for θ,

$$\arcsin x = \theta$$

(Remember arcsin, the inverse of sin? You probably thought it was just a dusty knickknack cluttering up the shelves in the beginning of the book, but lo and behold, it proves its worth after all.)
So

$$\int \frac{1}{\sqrt{1-x^2}} \, dx = \int d\theta$$

$$= \theta + C$$

$$= \arcsin x + C$$

Now let's try a harder one, where there are constants and other riffraff floating around.

Example *Calculate*

$$\int \frac{1}{4 + x^2}\, dx$$

Looks easier, doesn't it? In fact it's hard to believe you can't do it by regular u substitution, but you can't. Try it. Instead, we'll try to get $4 + x^2$ to look like one of the trig identities. Which one will do the trick? If we replace x with $\tan \theta$, we'll be looking at $4 + \tan^2 \theta$, which isn't quite what we want. Instead, let

$$x = 2 \tan \theta$$

Then

$$4 + x^2 = 4 + 4 \tan^2 \theta$$
$$= (4)(1 + \tan^2 \theta)$$

But hey, we just saw the trig identity $1 + \tan^2 \theta = \sec^2 \theta$. So

$$4 + x^2 = 4 \sec^2 \theta$$

Now mop up the dx. Since $x = 2 \tan \theta$,

$$\frac{dx}{d\theta} = 2 \sec^2 \theta$$

and

$$dx = 2 \sec^2 \theta \, d\theta$$

The integral becomes:

$$\int \frac{1}{4 + x^2}\, dx = \int \frac{1}{4 \sec^2 \theta} 2 \sec^2 \theta \, d\theta$$

$$= \int \frac{1}{2} d\theta$$

$$= \frac{1}{2}\theta + C$$

Remembering where we shelved x, we can solve for θ:

$$x = 2 \tan \theta$$

so

$$\arctan\left(\frac{x}{2}\right) = \theta$$

and we ultimately obtain the eminently satisfying

$$\int \frac{1}{4 + x^2}\, dx = \frac{1}{2} \arctan\left(\frac{x}{2}\right) + C$$

These types of problems are hard to create. So it's worth paying particularly close attention to the examples your instructor does in class and the examples you have to do for homework. Chances are good that when your instructor sits down at 4:00 A.M. to make up the next morning's exam, he or she will fall back on one of the examples you've seen in class with a slight change of constants.

28.3 Integration method: Partial fractions

You meet someone and really hit it off. You decide this is the big one, the person you want to spend your life with. The feeling is mutual and before you know it, there you are living together under the same roof. For a few weeks you are joined in blissful cohabitation. Every look fills your heart with tenderness. Then the relationship starts to change, begins to go sour. It's just little things at first. Your partner reads with a flashlight even when the lights are on and objects to your eating celery because the crunch is its cry of anguish. And then you catch your cuddle monkey cleaning the toilet with your toothbrush. Time to pull the rip cord.

This is the heart of the method of partial fractions. Sometimes, a pair of functions living under the single roof of the denominator just can't be integrated as is. No matter what method you try, they aren't getting along. They're intransigent. Time to pull the cord. Separate the two functions into two different fractions and blam, just like that, all your problems are solved. Both fractions can be easily integrated.

For instance,

$$\int \frac{3x - 1}{x^2 + x - 2}\, dx$$

is hard to integrate. But if we happen to notice that

$$\frac{3x - 1}{x^2 + x - 2} = \frac{\tfrac{2}{3}}{x - 1} + \frac{\tfrac{7}{3}}{x + 2}$$

we're all set. (You can check that this is true by combining the two fractions on the right-hand side over a common denominator.) Then we just have to integrate

$$\int \frac{\tfrac{2}{3}}{x - 1} + \frac{\tfrac{7}{3}}{x + 2}\, dx = \frac{2}{3}\ln|x - 1| + \frac{7}{3}\ln|x + 2| + C$$

But how did we know we could rewrite

$$\frac{3x - 1}{x^2 + x - 2}$$

as

$$\frac{\tfrac{2}{3}}{x - 1} + \frac{\tfrac{7}{3}}{x + 2}$$

That is some tricky algebra! And that's what the method of partial fractions tells us, the tricky algebra part. Here's how it goes. First we factor the denominator:

$$x^2 + x - 2 = (x - 1)(x + 2)$$

Then we write

$$\frac{3x - 1}{x^2 + x - 2} = \frac{A}{x - 1} + \frac{B}{x + 2}$$

A and B have yet to be determined. But notice that if we do get a common denominator for the two fractions on the right, it will be the same as the denominator on the left, a good sign. Now, we will just figure out what A and B should be in order for the left side of the equation to equal the right side.

Multiplying both sides by $(x - 1)(x + 2)$, we have

$$3x - 1 = A(x + 2) + B(x - 1)$$

Multiplying through,

$$3x - 1 = Ax + 2A + Bx - B$$

Gathering terms gives

$$(3 - A - B)x - 2A + B - 1 = 0$$

Since x can be anything, we see that when $x = 0$, we have

$$-2A + B - 1 = 0$$

But then we are left with

$$(3 - A - B)x = 0$$

and choosing $x = 1$ (since the equation has to hold for any x), we have

$$A + B = 3$$

Therefore, $-2A + B - 1 = 0$ and $A + B = 3$ both hold.

Solving the first for B and substituting into the second, we have $3A + 1 = 3$ and so $A = 2/3$. Then $B = 7/3$. So the reward for all this work is that we can now write

$$\frac{3x - 1}{x^2 + x - 2} = \frac{2/3}{x - 1} + \frac{7/3}{x + 2}$$

We can easily integrate these pieces, using substitution and the fact that $\int 1/x \, dx = \ln|x| + C$.

For students shooting for an A+: If the integrand is one polynomial divided by another, such as $\dfrac{x^3 - x^2 - 7x + 2}{x^2 - 3x + 2}$, and the highest power in the numerator is bigger than the highest power in the denominator, divide the bottom polynomial into the top, using long division of polynomials before you use partial fractions to integrate the remainder term. So in our case, dividing $x^2 - 3x + 2$ into $x^3 - x^2 - 7x + 2$ we obtain $x + 2 + \dfrac{-3x - 2}{x^2 - 3x + 2}$. We can directly integrate $x + 2$, and for the second part we just apply partial fractions to it and then integrate. Piece of cake.

Twenty Most Common Exam Mistakes

Here we will list some of the most common mistakes we see on exams. If you can avoid these, then at least your mistakes will be uncommon. Most of the mistakes that occur repeatedly involve algebra rather than calculus. They can be avoided by being careful and checking your work. Others involve common misunderstandings about various aspects of calculus.

1. $(x + y)^2 = x^2 + y^2$ MISTAKE!

Powers don't behave this way. The correct way to expand this expression gives

$$(x + y)^2 = x^2 + 2xy + y^2$$

2. $\dfrac{1}{x + y} = \dfrac{1}{x} + \dfrac{1}{y}$ MISTAKE!

The rule for adding fractions gives $\dfrac{1}{x} + \dfrac{1}{y} = \dfrac{x + y}{xy}$.

3. $\dfrac{1}{x+y} = \dfrac{1}{x} + y$ MISTAKE!

This very common error comes from carelessness about what's in the denominator. Can be avoided by careful handwriting or frequent use of parentheses.

4. $\sqrt{x+y} = \sqrt{x} + \sqrt{y}$ MISTAKE!

There is no simplified way to write $\sqrt{x+y}$. You just have to live with it as is.

5. $x < y$ so $kx < ky$, where k is a constant   MISTAKE!

This is true when k is a *positive* constant. If k is negative you need to reverse the inequality. If k is zero all bets are off. For example, if $x < y$ then $-x > -y$.

6. Forgetting to simplify fractions in limits   MISTAKE!

It is not correct to say $\displaystyle\lim_{x \to 1} \dfrac{x^2 - 1}{x - 1} = \dfrac{0}{0}$ and therefore the limit is undefined.

Even worse would be to cancel the zeroes and say the limit equals one.

Any time you get 0/0 for a limit, it is a BIG WARNING SIGN that says YOU HAVE MORE WORK TO DO! In this case,

$$\lim_{x \to 1} \frac{x^2 - 1}{x - 1} = \lim_{x \to 1} \frac{(x-1)(x+1)}{x-1} = \lim_{x \to 1} (x + 1) = 2$$

7. $\dfrac{\sin 2x}{x} = \sin 2$ MISTAKE!

You can cancel terms in the numerator and denominator of a fraction only if they are not inside anything else and are just multiplying the rest of the numerator and denominator. The function $\sin 2x$ is *not* $\sin 2$ multiplied by x. If the fraction had been written as

$$\frac{\sin(2x)}{x}$$

it would be harder to make such an error.

8. $ax = bx$ therefore $a = b$ MISTAKE!

This is a more subtle mistake. The cancellation is correct if x is not 0. For example, $2x = 3x$ forces $x = 0$. You cannot cancel the x and conclude that $2 = 3$. Not in this universe, anyway.

9. $\dfrac{d}{dx} 2^x = x\,2^{x-1}$ MISTAKE!

The correct answer is $2^x (\ln 2)$. The power rule only applies if the base is a variable and the exponent is a constant, as in x^3.

10. $\dfrac{d}{dx} \sin(x^2 + 1) = \cos 2x$ MISTAKE!

This is a typical example of the kind of mistakes made when applying the chain rule. The correct answer is

$$\frac{d}{dx} \sin(x^2 + 1) = (\cos(x^2 + 1))2x$$

11. $\dfrac{d}{dx} \sin(x^2 + 1) = \cos(x^2 + 1) + \sin 2x$ MISTAKE!

Another common way in which the chain rule is misapplied. This time the product rule has been used in a setting where the chain rule was the way to go.

12. $\dfrac{d}{dx} \cos x = \sin x$ MISTAKE!

The answer should be $-\sin x$. Extremely common error costing students over 10 million points a year on exams around the world.

13. $\dfrac{d}{dx}\left(\dfrac{f}{g}\right) = \dfrac{fg' - gf'}{g^2}$ MISTAKE!

This is backward! It should be

$$\frac{d}{dx}\left(\frac{f}{g}\right) = \frac{gf' - fg'}{g^2}$$

14. $\dfrac{d}{dx}(\ln 3) = \frac{1}{3}$ MISTAKE!

The quantity $\ln 3$ is a constant, so $\dfrac{d}{dx}(\ln 3) = 0$. The same is true for *all* constants. So $\dfrac{d}{dx}(e) = 0$ and $\dfrac{d}{dx}\left(\sin \dfrac{\pi}{2}\right) = 0$ as well.

15. $\displaystyle\int x \, dx = \dfrac{x^2}{2}$ MISTAKE!

The correct answer is $\displaystyle\int x \, dx = \dfrac{x^2}{2} + C$. Picky profs penalize points pedantically.

16. $\displaystyle\int \dfrac{1}{x} \, dx = \dfrac{x^0}{0} + C$ MISTAKE!

The power rule for integration does not apply to x^{-1}. Instead,

$$\int \frac{1}{x} dx = \ln|x| + C$$

17. $\int \tan x \; dx \;=\; \sec^2 x + C$ MISTAKE!

It's the other way around. $\dfrac{d}{dx} \tan x \;=\; \sec^2 x$. The correct answer is

$$\int \tan x \; dx \;=\; \ln|\sec x| + C$$

as can be found by u substitution with $u \;=\; \cos x$.

18. Forgetting to simplify   MISTAKE!
For example,

$$\int x \sqrt{x} \; dx$$

is easy if you notice that $x \sqrt{x} \;=\; x^{3/2}$ and then apply the power rule for integration. But if you try to do it using integration by parts or substitution, you will find yourself in outer space without a space suit.

19. Not substituting back to the original variable   MISTAKE!

$$\int 2x e^{x^2} \; dx$$

does not equal $e^u + C$ (it equals $e^{x^2} + C$).

20. Misreading the problem   MISTAKE!
If asked to find an area, don't find a volume. If asked to find a derivative, don't find an integral. If asked to use calculus to solve a problem, don't do it in your head using algebra. Although it seems silly to include this item in our list, billions of points have been taken off exams for mistakes of this type. After you finish a problem on the exam, go back and read the question again. Check to make sure you answered the question that was asked.

21. Thinking you're prepared when you're not   MISTAKE!
This mistake is perhaps the most important, so we'll put it in even though it pushes us over the 20 mistakes limit. The worst mistake many students make is to think they know the material better than they really do. It's easy to fool yourself into thinking you can solve a problem when you're looking at the answer book or at a worked-out solution. Test your knowledge by trying problems under exam conditions with the book closed, the TV off, and the friend who loves to point out your misplaced minus signs locked out of the room. If you can do them under these restrictions, the exam should be a breeze.

What's on the Final?

Ah, the final. That cumulative exercise that gives you the opportunity to regurgitate the material from an entire course onto a couple of pieces of paper within a very short time span. Like an oncoming storm spied on the horizon, it strikes fear into the entire student body. It has the power to destroy budding academic careers, to annihilate entire fraternities, to bring beef-bound athletes to their knees. But don't worry. Because in this chapter, we tell you exactly what we think will be on the final, at least probabilistically speaking. And then you will be prepared, so you can nail that sucker to the wall.

Here is a list of types of problems and how likely they are to be on the exam. Of course, your teacher may have his or her own favorite types of problems and so your final may diverge from this list. The best source of information is always your own instructor. (Whew, now you can't sue us if the types of problems on your final don't look anything like ours.)

1. Curve-sketching problems (most common; occur on 99% of final exams). These problems ask you to sketch a graph. They might say something like:

Sketch the graph of the following function. Indicate its critical points, asymptotes, and intercepts. Determine inflection points and show where it is concave up and where it is concave down.

This often involves nasty functions like

$$f(x) = 3x^{2/3} - x^2 + 1$$

or

$$g(x) = \frac{1}{x} + \frac{4}{x^3}$$

The idea is that you will use your knowledge of calculus to find maxima, minima, etc. That is all okay, but plotting a few points on the side can help to give the general plan and check your other calculations. A good place to start on these problems is to calculate the intercepts by setting x and y, respectively, equal to 0. Are there points where the function is undefined—say, where a denominator becomes 0? Maxima and minima can be found by differentiating and setting the derivative equal to 0. These problems are usually set up so this is easy to do. Remember to plug the critical points into the function to get the critical values. You need an (x, y) coordinate to put on your graph. Making a table will help you keep your x and y coordinates straight.

2. Differentiation problems (most common; occur on 100% of final exams). These involve differentiating functions, including some possibly unattractive ones. These functions often need to be split apart into easy-to-do components using the chain rule, product rule, etc. Here are some examples:

Find $f(\dot{x})$ if

$$f(x) = \ln(\sin x)$$

$$f(x) = x(3x - x^2)^6$$

$$f(x) = \sec x$$

Write everything down carefully and use lots of parentheses. That way you avoid making algebraic mistakes, one of the chief causes of trouble on exams.

3. Limit problems (very common; occur on 90% of final exams). A typical limit problem is

Find each of the following limits:

$$\lim_{x \to 2} f(x)$$

$$\lim_{x \to 2^-} f(x)$$

$$\lim_{x \to 2^+} f(x)$$

where

$$f(x) = \begin{cases} -1 & \text{if } x \le 1 \\ 2x & \text{if } 1 < x < 2 \\ x^2 + 1 & \text{if } 2 \le x \end{cases}$$

In this problem the function is defined differently on different intervals. Perhaps because of some deep rooted psychological trauma at birth, this kind of thing seems to terrify most students. Probably it's just a lack of practice with the notation. A good way to see what's going on is to sketch the function.

Another common type of limit problem is

$$\lim_{x \to 2} \frac{x - 2}{\sqrt{x + 2} - 2}$$

Both numerator and denominator are approaching 0 as $x \to 2$. Not good. But this problem can be solved by some algebra that cancels the common 0:

$$\frac{x - 2}{\sqrt{x + 2} - 2} = \frac{x - 2}{\sqrt{x + 2} - 2} \frac{\sqrt{x + 2} + 2}{\sqrt{x + 2} + 2}$$

$$= \frac{(x - 2)(\sqrt{x + 2} + 2)}{x + 2 - 4}$$

$$= \frac{(x - 2)(\sqrt{x + 2} + 2)}{x - 2}$$

$$= \frac{\sqrt{x + 2} + 2}{1}$$

and the answer can now be found by judicious plugging in:

$$\lim_{x \to 2} \sqrt{x + 2} + 2 = \sqrt{2 + 2} + 2 = 4$$

The point to remember: When plugging in a value gives $0/0$ or ∞/∞, try doing some cancellation. (If that doesn't work, you may need a more sophisticated approach such as L'Hôpital's rule. If L'Hôpital's rule doesn't ring any bells, then you won't be needing it. If it does ring a bell, go see who's at the door.)

A hard limit problem might also involve knowing that $\lim\limits_{x \to 0} \dfrac{\sin x}{x} = 1$.

4. Max/min word problems (very common; occur on 95% of final exams). Typically, these problems ask you to translate a word problem into mathematical formulas and then find the maximum or minimum value of a function. Often the hardest part of these problems is to set up a function to differentiate. Finding the max or min is done by differentiating, setting the derivative equal to zero, and solving for the critical points. You then use such tests as the second derivative test, the first derivative test, or the plug-it-in test to see if you are at a maximum, minimum, or neither.

   Common pitfalls: The maximum may occur at the boundary of an interval, in which case the derivative may not be zero at the absolute maximum.

5. Domain/continuity problems (common; occur on 50% of final exams). These problems ask you to determine the domain of a function and find where the function is continuous. The function might be defined in pieces, like

$$f(x) = \begin{cases} -1 & \text{if } x \le -1 \\ \dfrac{1}{x} & \text{if } -1 < x < 2 \\ x^2 - 1 & \text{if } 2 \le x \end{cases}$$

The things to look for are points where a denominator becomes zero, such as $x = 0$ in our function, and what happens at one of the points where the function's formula changes, occurring at the points -1 and 2 for our $f(x)$.

6. Limit definition of the derivative (common; occur on 50% of final exams). These problems ask you to compute a derivative directly from the limit definition. There are not too many functions for which this is feasible. Here is a list of likely candidates:

   (a) x, x^2, or x^3. Higher powers are unlikely.

   (b) $\sqrt{x}$. The limit can be done by multiplying top and bottom by $\sqrt{x + h} + \sqrt{x}$.

   (c) $1/x$

   (d) $(x - 1)/(2x + 5)$ and similar simple quotients of polynomials.

7. Speed/velocity problems (common; occur on 50% of final exams). Typically these problems involve baseballs, race cars, bullets, airplanes, yo-yos, or

sky divers (with or without parachutes). You will need to remember that velocity is the derivative of position, and acceleration the derivative of velocity. Keep your signs straight. Remember that what goes up must come down, what goes around comes around, and you are what you eat.

8. **Related rates** (common; occur on 50% of final exams). Mickey and Minnie Mouse are an example. Oops! They're related rats. Related rates problems are just what they sound like—two rates of change that are related. You can express one in terms of the other, and that's just what you have to do to solve these problems. Usually the hardest part is to translate the problem into equations and to figure out what you're being asked for.

9. **Implicit differentiation** (common; occur on 40% of final exams). If you can't solve the equation for y, you are going to have to use implicit differentiation to compute dy/dx. If you are asked to compute dy/dx at a point (x_0, y_0), with both coordinates given to you, this is a tip-off to use implicit differentiation.

10. **Differentials** (less common; occur on 20% of final exams). The question will ask you to estimate something, perhaps $\sqrt{3.9}$ or $\sin(\pi + 0.1)$.

11. **Equations of tangent lines** (less common; occur on 20% of final exams). These problems ask you to find the equations of tangent lines. The trick is that you need to differentiate to determine the slope of the line.

12. **Compute the integral problems** (most common; occur on 100% of final exams). Generally, the exam presents a section of integrals where you are not told what method to use and you have to figure it out on your own. Here's a quick sample list with method:

* Substitution: $\int x^2 e^{x^3}\,dx,\ \int x\sqrt{x^2+1}\,dx,\ \int \sin^3 x\cos x\,dx$

* Integration by parts: $\int \ln x\,dx,\ \int x\ln x\,dx,\ \int xe^x\,dx,\ \int x\sin x\,dx,$ $\int e^x\sin x\,dx$ (This one is tricky. You need to use parts twice and rearrange terms.)

* Partial fractions: $\int \dfrac{1}{x^2-1}\,dx,\ \int \dfrac{2x-3}{x^2-5+6}\,dx$

* Trigonometric substitution: $\int \dfrac{1}{x^2+1}\,dx,\ \int \dfrac{1}{\sqrt{1-x^2}}\,dx$

13. **Find the area between curves** in the plane (common; occur on 50% of final exams). Here you take the top function minus the bottom function as your integrand and integrate between either the given limits or the limits determined by where the two graphs intersect.

14. Numerically approximate a definite integral (less common; occur on 20% of final exams). Here you use a finite Riemann sum, the rectangle rule, the trapezoid rule, or Simpson's rule to approximate a definite integral. Not asked much because they're time consuming, measure little but memorization skill, are messy for students to work out, and are painful for instructors to grade.

15. Exponential growth and decay problems (common; occur on 70% of final exams). If these problems were covered in the course, they are most likely going to be on the final.

GLOSSARY: A Quick Guide to the Mathematical Jargon

Absolute Maximum   The all-time, one-and-only, single, absolute, and total maximum value of a function over a specified domain of the function. (Although it is the unique maximum value, it could occur at more than one point, as when you have two mountain peaks of exactly the same height.) Not to be confused with a local maximum, which is to the absolute maximum as the police chief is to the attorney general. The absolute maximum is sometimes also called the global maximum. Remember to check whether it occurs at the endpoints of the interval!

Absolute Minimum   Same definition as for absolute maximum, but substitute the word "minimum" everywhere the word "maximum" occurs. Also substitute "precinct boss" for "police chief" and (optionally, depending on your politics) "president" for "attorney general." On a graph it is a point that is having a really bad day—as low as it can get.

Absolute Value   Drop the negative sign if there is one. Otherwise, just leave the number alone.

Acceleration   Rate of change of the velocity. It causes that funny feeling in the pit of your stomach as you are mashed backward into the seat when somebody really puts the pedal to the metal. Since the rate of change of a function is its derivative, the acceleration is the derivative of the velocity function. Since the velocity function is itself the derivative of the function giving your position, the acceleration function is the second derivative of the function giving your position. (In mathese, $a = dv/dt = d^2s/dt^2$, where s is the position function.)

Algebra   Hold it. If you don't know what algebra is (a bunch of letters like x and y and a bunch of rules for playing around with them), then you shouldn't be taking calculus. Return to GO; do not collect $200.

Antiderivative   You guessed it. This is the opposite of the derivative. Doesn't deserve the negative connotations associated to some of the other "anti" words like "antichrist," "antisocial," or "antimacassar" (that little lace doily you used to see on your grandma's couch, made obsolete by plastic slipcovers). The antiderivative of a function $f(x)$ is another function $F(x)$ whose derivative is $f(x)$. Also called the *indefinite integral* of $f(x)$. The "antiderivative" terminology is traditionally usually used just before the introduction of indefinite integrals, and then never used again, having been forever replaced with the term "indefinite integral."

213

Antidifferentiation Process of taking an antiderivative. Also, a strong aversion to distinguishing among different people, as with parents who insist on calling all five of their children "Frank."

Asymptote An asymptote is like one of those people you meet at a party who is devastatingly attractive and you just want to get close. You maneuver your way next to him or her and casually strike up a conversation. Making good time, you get closer and closer, till you're practically knocking knees. In calculus, you just keep getting closer. In the real world, you start explaining your love of partial fractions, he or she excuses himself or herself to get a drink, and then you see a car driving away through the window.

An asymptote for the graph of a function is a line sitting in the xy-plane that the graph of the function approaches, getting closer and closer as we travel along the line. Functions that have had one too many may weave back and forth across an asymptote, but still, the further out you go, the closer they get.

Callipygian Appears near "calculus" in the dictionary. Look it up.

Carbon Dating Essence of the social life of geologists. They get together, crush a bunch of rocks, and then determine the amounts of various types of carbon in the rock. Since carbon-12 does not decay over time and carbon-14 does decay over time, they can tell by the ratio of carbon-14 to carbon-12 how old the rocks are. What is this topic doing in a calculus book? The rate of decay of the carbon-14 and any other radioactive substance is exponential. That is to say, the amount at time t is given by $f(t) = C_0 e^{-kt}$. A great source of problems and examples.

Cartesian Coordinates Standard coordinates in the plane—you know, the ones where you have an x-axis and a y-axis, and each point is given by specifying two numbers (7, 4), which means go out 7 units in the x direction and then 4 units in the y direction. Why the funny name? They are named after the French mathematician René Descartes, whose Latin name was Cartesius.

Cartesian Plane Plane upon which are Cartesian coordinates. It also describes the entire air force of the country of Cartesia.

Chain Rule "Never allow yourself to be chained up by someone whose body is covered by more tattoos than latex." The mathematical version states

$$(f(g(x))' = f'(g(x))g'(x)$$

or

$$\frac{df}{dx} = \frac{df}{du}\frac{du}{dx}$$

Completing the Square   Here's a phrase that gets thrown around a lot and is the kind of thing that every teacher assumes some other teacher has shown you before. It's best demonstrated by example. If we want to complete the square on $x^2 + 8x + 10$, we write it as

$$x^2 + 8x + 10 = x^2 + 8x + (8/2)^2 + 10 - (8/2)^2$$

$$= x^2 + 8x + (8/2)^2 - 6 = (x + 4)^2 - 6$$

Why would we want to complete the square on a quantity? For one example, suppose that you want to graph $x^2 + 8x + 10 + y^2 = 0$. By completing the square, this becomes

$$(x + 4)^2 + y^2 = 6$$

This is the equation of a circle of radius $\sqrt{6}$ centered at the point $(-4, 0)$.

Completing the Square Dance   "We're not playing music anymore, so swing your butts right out the door."

Complex Number   Number that neglected to "get real." Currently in therapy. Also, one of those numbers like $7 + 6i$, where i is the number $\sqrt{-1}$. We know, everybody says you can't take the square root of a negative number, but what they really mean is that you can't take the square root of a negative number and expect to get a real number. Which is true—you get a complex number instead. Given a complex number of the form $a + bi$, a is called the real part and bi is called the imaginary part. Normally doesn't come up in a first calculus course.

Composition of Functions   Applying one function to another. For instance, $\sin(\sqrt{x})$ is the composition of $\sin x$ with $\sqrt{x}$. If successful, the two functions are then performed by an orchestra.

Concavity   A part of the graph of a function is said to be concave down if it looks like part of a frown and concave up if it looks like part of a cup (... up ... cup, there's a mnemonic device). In order to tell whether a function is concave up or down, one uses the infamous second derivative test, which is discussed in detail in Sections 15.3 and 15.4.

Constant   Fixed number, like 3 or $\sqrt{2}$. To be distinguished from a variable, which has no single value. When you say, "My spouse was my constant supporter," you mean that he or she never wavered, despite your conviction for arms dealing and tax evasion, your decision to come out of the closet, and your involvement in the Perot for President campaign.

Continuity   You know, no big surprises. Everything keeps going forward on an even keel. Here's the technical definition: A function $f(x)$ is continuous at a point a if

$$\lim_{x \to a} f(x) = f(a)$$

Moreover, a function is said to be continuous if it is at every point where it is defined. Now for a less technical definition: A function is continuous everywhere if you can draw the entire graph of the function without lifting your pencil from the page. (Okay, you can lift your pencil long enough to draw the axes.) See Chapter 9 for more details.

Critical Point   Point that was made when you weren't paying attention. Also, a value of x that makes the derivative $f'(x)$ of a function either equal to 0 or nonexistent. It comes up either in graphing functions, telling you where the critical changes in the graph occur, or in applied max-min problems, where it tells where the potential maxima or minima are occurring.

Definite Integral   The definite integral of a function $f(x)$ over an interval $a \leq x \leq b$ is a number, sometimes thought of as the area under a graph. Not to be confused with the indefinite integral, which gives a function.

Derivative   Hey, this is the most important idea in all of calculus. You shouldn't be looking up the definition as if it's some word in the dictionary like "apothecary." You should be reading about it in this book (see Chapters 10, 11, and 12). But if you insist on a nutshell definition, the derivative of $f(x)$ is the rate of change of $f(x)$. Geometrically, it also represents the slope of the tangent line to the graph of the function $y = f(x)$ at the point $(x, f(x))$, but that's a mouthful.

Dictionary Table Tennis   See **Lexicon Ping-Pong.**

Differentiable Function   A function is differentiable at a point if its derivative exists at that point. For instance, $f(x) = x^2$ is differentiable everywhere, whereas $g(x) = |x|$ is differentiable everywhere except $x = 0$. Why don't we say "derivativable"? Because it sounds ridiculous.

Differential   It has something to do with the transmission of your car, but it's way too complicated for us to understand. Oh, yeah, a differential is also a small change in a variable. For instance, dy is a small change in the value of y. Although dy often occurs as part of the symbol for the derivative and as part of the symbol for the integral, and although in those other guises dy plays a role very similar to the one intended when we call it a differential, it is best to just think of the differential dy as a very small change in y.

Differential Equation   An equation that involves derivatives, as in

$$6\frac{dy}{dx} + y - x = 25$$

These equations govern most of the physical world, so treat them with respect.

Domain of a Function   A dog's domain is all of the land that he can traverse in a day, starting with one full bladder. A function's domain is just the set of all values for x that it makes sense to plug into $f(x)$. For instance, the domain of $f(x) = \sqrt{x}$ is all $x \geq 0$.

Double Integration   Calm down. It's okay. If you are looking up this word, then that means some jerk from multivariable calculus has said to you, "If you think integration's hard, wait until you hit double integration." First of all, it's a lie. Double integration isn't that hard. And second, you don't need to worry about it for quite a while yet. Back to the good stuff.

e e is one of those numbers that is so important, it gets its very own name. In fact, $e = 2.71828 \ldots$. Why is it so important? Have you ever tried to write a sentence without it? It comes up all over the place. In fact it's the most commonly occurring letter in the whole alphabet! Same thing in calculus. One tantalizing tidbit is that it is the *only* number you could pick such that

$$\frac{d}{dx}e^x = e^x$$

Ellipse   Step on a circle until it squawks, and you've got an ellipse. It's a bit longer than it is wide, or vice versa. The general formula is like the formula for a circle with a few extra a's and b's thrown in:

$$\frac{x^2}{a^2} + \frac{y^2}{b^2} = 1$$

Exponent   That little number that appears as a superscript next to another number or function. Also called the power. If you are divorced, this is not what people are referring to when they say, "How's your ex?"

Exponential Function $f(x) = e^x$. Its most famous property? It is its own derivative. That's like being your own mother, not so easy to do.

Exponential Growth   Exponential growth is VERY VERY FAST GROWTH. When people say exponential growth, they are trying to impress the hell out of you. A function experiences exponential growth if it is at least as big as a function of the form CK^x, where $C > 0$ and $K > 1$. For example, the function 2^x experiences exponential growth. Notice that the function 2^x doubles in value each time x increases by one. So although 2^1 is only 2, 2^{10} is already 1024, and 2^{20} is 1,048,576. That is VERY VERY FAST GROWTH.

Extremum (plural, **Extrema**) Just a word for either a maximum or a minimum. Let's face it: A maximum or a minimum is a point where a high or low extreme occurs.

Factorial $n! = n \times (n-1) \times \cdots \times 3 \times 2 \times 1$, that is to say, the factorial of an integer n is simply the product of all of the integers from 1 to n. Here's a good question to stump your professor with: "How do you take the factorial of a number like $3/2$ or -2?"

Function Functions are something that just about everyone encounters during the four years of college, even the TV majors who avoid any class where the word "calculate" is used. There are two types of functions—social functions and mathematical functions. Though completely different, they use much of the same terminology.

Social functions are also called mixers or gatherings. Usually they involve parties hosted by dormitory floors (or assistant deans) with kegs of beer (or little cucumber sandwiches). The location where the function takes place is known as the domain of the function. The place where the food is cooked is known as the range of the function. A function that lasts until morning is said to be continuous. One that is broken up by the police and resumed the next day is called discontinuous. The phone number of the dreamboat you met is called the value of the function. It often winds up in the range.

The same terminology is used by mathematicians to describe what they call a function. The main difference is that when a mathematician has a function, everyone gets exactly one value! No one leaves with two numbers at a mathematical function, and no one leaves with none.

A mathematical function is a machine where you put in a real number (often denoted by a variable x but sometimes by t or some other letter) and it spits out a new real number. For instance, $f(x) = x^2$. You put in the number 3 for x and it spits out the number 9. Its domain is the set of values that are legal to put in, and its range is the set of possible values it can spit out.

Fundamental Theorem of Calculus This theorem is usually stated in two parts. One part states that finding areas under curves can be done by taking antiderivatives and plugging in the limits.

$$\int_a^b f(x)\, dx = F(b) - F(a)$$

where $F(x)$ is any function whose derivative is $f(x)$, that is to say, $F'(x) = f(x)$. This can also be stated as

$$\int_a^b F'(x)\, dx = F(b) - F(a)$$

If you integrate the derivative of a function over an interval from a to b, you just get the original function evaluated at b minus the original function evaluated at a. The other part states that

$$\frac{d}{dx}\int_{a}^{x} f(t)\,dt = f(x)$$

Both parts show that derivatives and integrals are intimately related, and we don't just mean on a first-name basis. If it weren't for this theorem, calculus courses would be half as long as they are.

Global (Extremum, Maximum, Minimum, Warning) Another expression for the absolute extremum, maximum or minimum. It comes from the fact that this extremum is the most extreme extremum on the globe. For instance, the coldest place on earth, emotionally speaking, is Washington, D.C. It is a global extremum.

Graph Pictorial representation of a function formed by plotting $f(x)$ in the y direction on the xy-plane. Very useful, because while pictures say a thousand words, a graph gives an infinite number of function values.

Given a function $f(x)$, the graph of the function is simply the set of points (x, y) in the Cartesian plane that satisfy the equation $y = f(x)$. Most important property? Any vertical line can intersect the graph at most once, since a vertical line is defined by a particular value of x. But for a particular value of x, there is only one value of y such that $y = f(x)$.

Hyperbolic Trigonometric Functions Well, this means you are in a slightly more heavy-duty calculus course. Most calculus courses skip this material because it is expendable and they run out of time. But your course isn't skipping it. That's okay, because these functions are actually a cinch to deal with. The hyperbolic sine of x, denoted $\sinh x$ (pronounced like cinch— and it is) is defined to be $\sinh x = \dfrac{e^{x} - e^{-x}}{2}$ while the hyperbolic cosine, denoted $\cosh x$ (rhymes with posh), is defined to be $\cosh x = \dfrac{e^{x} + e^{-x}}{2}$. Note that each is the other's derivative. All of the other hyperbolic trig functions are defined in terms of these two in exactly the same way the other trig functions are defined in terms of the sine and cosine functions.

Hypocycloid Just kidding. We don't know, some kind of cycloid, probably. Shouldn't you be working some problems?

Indefinite Integral The indefinite integral of a function $f(x)$ is another function $F(x)$ with the distinguishing feature that the derivative of $F(x)$ is $f(x)$. Not to be confused with the definite integral, which gives a number as an answer. Also called the *antiderivative* of $f(x)$.

Integer   ..., $-3, -2, -1, 0, 1, 2, 3, ...$ How's that for a short definition?

Integrable   If people have integrity you can count on them. If people have integrability you can take their antiderivatives. Same holds for functions. A function is integrable if its integral exists. Most of the standard functions are integrable.

Integrand   Function inside the integral being integrated. Found between the $\int$ and the dx.

Integrandstand   Stand upon which to put the function inside the integral that is being integrated.

Inverse Trigonometric Function   An inverse trigonometric function is the function that reverses the effect of the original trig function, kind of like the Democrats and the Republicans, when they are taking turns being elected to power. The one undoes whatever the other had accomplished while in office. The inverse function for sine is denoted arcsin. So if $y = \sin x$, then $x = \arcsin y$. The notation arcsin is used, rather than $\sin^{-1}$, so as to prevent people from confusing the inverse trig function with $\dfrac{1}{\sin x}$.

Irrational Number   Number a few apples short of a picnic. Also, a real number that is not rational, which is to say that it cannot be written as a fraction of two integers. Classic examples include π, e, and $\sqrt{2}$. Every irrational number has a decimal representation that is nonrepeating. There are tons of these numbers, actually more of these than there are of the rational numbers. (Mathematicians say there are uncountably many of these.) Of course, you would be completely justified in noting that there are infinitely many rational numbers, so how could there possibly be more of these than there are of the rational numbers? That would bring us to the topic of the different kinds of infinity, but that's a little too far afield for us. Good question to ask the professor though.

Lexicon Ping-Pong   See **Dictionary Table Tennis.**

Limit   That bound you cannot exceed, as in "limit of three trips to the salad bar per customer." In calculus, a limit is the number you approach as you plug values into a function, and the values get closer and closer to a given number.

Line   What we hope you waited in to get a copy of this book.
      In math, there are not a whole lot of questions about what a line is. It's that straight thing between any two points. The equation of a line has two general

forms, the point-slope form $y - y_0 = m(x - x_0)$, where (x_0, y_0) is a point on the line and m is the slope of the line, and the slope-y-intercept form $y = mx + b$, where b is the y-coordinate where the line intersects the y-axis and m is the slope.

Linear Equation Equation that represents a line. Looks something like $3x + 2y = 4$. No x^2 or $\sin x$ or even an xy. It can always be put in a general form $Ax + By + C = 0$, where A, B, and C are constants that are possibly zero. (Note that both of the equations for lines in the previous definition can be put in this form.)

Local (Extremum, Maximum, Minimum) If you were nearsighted, this point would look like one where the graph of our function has an extremum, namely, a maximum or a minimum. Possibly if you expanded your vision you would see a larger or smaller value somewhere far away on the graph, but comparing this point only to its near neighbors, it comes out on top (in the case of local maximum) or on the bottom (in the case of local minimum). The local maximum is a big fish in a small pond. A local minimum may be able to find someone even lower than it if it wanders out of its neighborhood.

Logarithm The beat of trees being cut down in the Pacific Northwest. Also, a mathematical function that is the inverse of b^x for some fixed number called the base of the logarithm.

Map Danger! Danger! If your professor uses this word to refer to a function, then you are in serious trouble. That means that she is a theoretical mathematician and incapable of separating the theoretical world from the world of the classroom. "Map" is another word used for a function: "This is a map from the reals to the reals" translates as "This is a function that takes a real number and turns it into another real number."

Maxima Plural of "maximum." One of the few examples of an application of Latin. Who says the language is dead? It's just resting. See **Local Maximum** and **Absolute Maximum** for more details.

Minima Look, we just explained maxima. Do we have to do everything around here?

Natural Logarithm The natural logarithm, usually denoted ln, is the logarithm to the base e. Also, a method of birth control used at tree farms.

Negative Pessimistic or depressed. Often treatable with Prozac.

Origin The point in space with all coordinates equal to 0. Thought to be somewhere in Iowa.

Orthogonal   Fancy math word for "perpendicular." When you say, "He's orthogonal to the rest of the world," you mean he's perpendicular to everyone else, living in a slice of his mind that the rest of us don't have.

Parabola   Certain type of curve. An equation of the form $y = Ax^2 + Bx + C$ will always give a parabola. The most common example is $y = x^2$, a curve passing through the origin in the shape of an upward-opening cup. The prefix *para* comes from the Greek and means "at or to the side of," as in *paralegal:* at or to the side of a lawyer; *paranormal:* at or to the side of normal; *Paraguay:* at or to the side of Guay, a tiny country most people don't know about. Parabolas can also be obtained by slicing a right circular cone by a plane that is parallel to a line in the cone passing through the vertex. How about that for a useful fact?

π That leader in the Number Hall of Fame, $= 3.14159\ldots$. It can be defined to be one-half of the circumference of a circle of radius 1.

   You probably think that the letter π is used for the circumference of a unit diameter circle because it was thought up by some ancient Greeks. Well it was, but the letter π was not used for this number until a few hundred years ago, and this use was introduced by an Englishman at that. His motivation is unclear. Some suspect that it is because perimeter starts with a p. Others know that the English like a good pie for lunch.

   What some of those ancient Greeks did give us was the sorry method of measuring angles using 360 degrees. The origin of the number 360 is also murky, though it is suspected to have something to do with the fact that a pizza can be nicely divided into six slices. For calculations involving angles, it is much easier to work with a set of angles called radians. They give a good way of slicing the π, so to speak.

   For most purposes π is about 3.14. It took thousands of years before π was known to 10 decimal places. Today, mathematicians have calculated π to over 3 billion decimal places. Fortunately, most professors will not ask you to memorize more than the first hundred thousand.

π **à la mode.**   The number $3.14159\ldots$ with a big scoop of vanilla ice cream on top. Best when warmed.

Polynomial   You know, a function like $x^2 - 7x + 3$ or $2y^{15} - 4y^3 + 3y - 6$. Polynomials do not contain any square roots or trig functions or anything the slightest bit weird. In their general form, they look like

$$f(x) = a_n x^n + a_{n-1} x^{n-1} + a_{n-2} x^{n-2} + \cdots + a_2 x^2 + a_1 x + a_0$$

Position Function   Function that depends on time and tells you what your position is along a number line as time varies. For instance, if $f(t) = t^2$, in units of feet and seconds, then at time $t = 0$ you are at the origin, at time

$t = 1$ second you are 1 foot to the right of the origin, and at time $t = 2$ seconds you are 4 feet to the right of the origin. Of course at time $t = 52$ hours, you have traveled farther than the speed of light would allow, breaking one of the most basic laws of the physical universe. Cool.

Power Rule "Power corrupts. Absolute power corrupts absolutely" (where, by definition, absolute power equals power if power is positive, otherwise equals the negative of power). In calculus, the derivative of x^n equals nx^{n-1}.

Quadratic Formula That amazing formula for finding all values of x that satisfy the equation $ax^2 + bx + c = 0$. Works even if you can't factor the left-hand side. There are two solutions, which may be equal:

$$x = \frac{-b + \sqrt{b^2 - 4ac}}{2a}$$

or

$$x = \frac{-b - \sqrt{b^2 - 4ac}}{2a}$$

Also slices and dices.

Radian Measure of angle. One radian is $\dfrac{180°}{\pi}$, believe it or not. And one degree is $\dfrac{\pi}{180}$ radians.

Range How far you can throw a ball is the range of your pitching arm. The set of values a function can take is the range of the function. What mathematicians cook on is the range in the kitchen. Fancy poultry on a menu is free-range chicken. Home, home on the range . . . never mind.

Rate of Change This is the speed at which a function is changing. If the function is measuring your position, then your speed (as measured by your speedometer) is your rate of change. Another name for the rate of change of a function is the derivative of that function.

Rational Function Function that makes a lot of sense. Also, the ratio of two polynomials, like $\dfrac{x^2 - 2}{2x^3 + 1}$.

Rational Number Number that has both feet on the ground. It's all there. Also, a number of the form a/b, where a and b are integers. A few famous ones are $\frac{1}{2}$ and $\frac{3}{4}$. A less famous one is $\frac{337}{122}$. Each rational number has a decimal representation that is either terminating (consisting of finitely many decimal places) or repeating. Interestingly enough, there are a lot more irrational numbers than rational numbers. Another example of math imitating life.

Real Number   Numbers we usually deal with, including the integers, the fractions, and the irrational numbers like e and $\sqrt{2}$ that occur between the fractions. Each has a decimal representation. To be distinguished from imaginary numbers, which involve $\sqrt{-1}$.

Second Derivative   Derivative that comes after the first derivative and before the third derivative. Obtained by taking the derivative of a function twice in a row.

Secant Line   Jargon term for a line through two points in a curve. Take a curve, any curve. Then take two points on the curve. Connect them by a line. That is a secant line. Why isn't it just called a line? History. Most often secant lines come up in references to tangent lines, where you take a sequence of secant lines, fixing one of the points on the curve and moving the second point along the curve toward the first point. The sequence of secant lines that you obtain approaches the so-called tangent line. Since the tangent line has such a fancy name, people felt bad for the secant line and gave it a fancy name, too.

sine and cosine   Two things that mathematicians ask each other at parties: "What's your sine?" and "What's your cosine?"

Speed   Absolute value of your velocity. It's used when you don't care whether you backed into the wall at 30 mph (where velocity is -30) or drove into the wall forward at 30 mph (where velocity is 30); you just want to tell people you hit the wall going 30.

Speedometer   That little gizmo in your car that tells you how fast you are going. It is essentially a velocity function. If you look at it at a given time, it tells you your velocity (rate of change of your position function) at that time, assuming you are not backing up.

Tangent Line   Line that rubs up against and "kisses" a curve at a point, having the slope of the graph at that point. Now facing charges for sexual harassment.

Theorems and Proofs   A theorem is a claim on some subject, such as "The derivative of $\sin x$ is $\cos x$." A proof is a detailed, logical, completely convincing argument showing why a theorem is true. Learning the difference between what does and what does not constitute a proof is one of the most important things you can get out of a calculus course, though there is seldom time for a detailed discussion of this issue in a crowded curriculum. Good proofs should convince all reasonable people. Of course there are always those people you meet at parties who will say, "Wait a minute, what if a Martian was hypnotizing me while I heard the argument?"or "Isn't truth all relative anyway? Why is one truth better than another?" Fortunately, they rarely get second invitations.

Mathematicians are often kept off juries because of a belief by lawyers that they cannot understand the legal meaning of "proof beyond a reasonable doubt." If you want to avoid jury duty, point out that you learned about theorems and proofs in calculus. For similar reasons, lawyers are kept out of calculus classes because of a belief by mathematicians that they cannot understand the mathematical meaning of "proof." (If you are a lawyer or future lawyer, please don't sue us.) Mainly because mathematicians, like lawyers, fall into the use of jargon, theorems are also called corollaries, lemmas, and propositions.

Trigonometric Identity Any simple equation that relates various trig functions. The most famous and important is the classic $\sin^2 x + \cos^2 x = 1$, although there are also many less significant ones running around underfoot. Note that the classic immediately gives you others such as $\tan^2 x + 1 = \sec^2 x$ by dividing through by $\cos^2 x$.

Velocity Rate of change of position (only differing from speed in that it can be negative if you are moving left along the number line). It is obtained by taking the derivative of your position function.

Variable Single word used most often by nervous meteorologists. In math, a quantity that can vary. Often represented by a letter like x or y, since it does not have a specific fixed value, but rather can take on a whole set of different values.

Zeno The last entry in any dictionary of calculus terms. Zeno was a Greek philosopher and is best known for Zeno's paradox. He pointed out that for a runner to get from A to B, he or she must first traverse half the distance, and then half the remaining distance and then half the remaining distance ad infinitum. Since, clearly, the runner cannot perform infinitely many steps in a finite amount of time, motion is an impossibility and is therefore an illusion. So all the world is just a dream. Roll over and go back to sleep.

INDEX

Algebra:

* Factoring: $x^2 - y^2 = (x - y)(x + y)$

* The quadratic equation:
  $ax^2 + bx + c = 0$ has solution

$$x = \frac{-b \pm \sqrt{b^2 - 4ac}}{2a}$$

  if $a \neq 0$.

* The absolute value function:

$$|x| = \begin{cases} x & \text{if } x \geq 0 \\ -x & \text{if } x < 0 \end{cases}$$

Trigonometry:

* The Trig. Functions

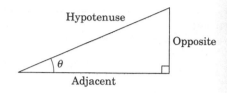

$$\sin \theta = \frac{\text{opposite}}{\text{hypotenuse}}$$

$$\cos \theta = \frac{\text{adjacent}}{\text{hypotenuse}}$$

$$\tan \theta = \frac{\text{opposite}}{\text{adjacent}} = \frac{\sin \theta}{\cos \theta}$$

$$\csc \theta = \frac{\text{hypotenuse}}{\text{opposite}} = \frac{1}{\sin \theta}$$

$$\sec \theta = \frac{\text{hypotenuse}}{\text{adjacent}} = \frac{1}{\cos \theta}$$

$$\cot \theta = \frac{\text{adjacent}}{\text{opposite}} = \frac{1}{\tan \theta}$$

* Degrees and Radians:

$$180° = \pi \text{ radians}; \quad 360° = 2\pi \text{ radians}$$

$$1° = \frac{\pi}{180} \text{ radians}, \quad x° = \frac{x\pi}{180} \text{ radians}$$

$$1 \text{ radian} = \frac{180°}{\pi}, \quad x \text{ radians} = \frac{180x}{\pi}°$$

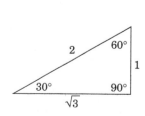

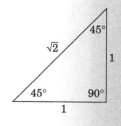

How to Ace Calculus: Just the Facts

Important Values of Trig Functions:

(* means the value doesn't exist)

degrees	0°	30°	45°	60°	90°	180°	270°
radians	0	$\pi/6$	$\pi/4$	$\pi/3$	$\pi/2$	π	$3\pi/2$
$\sin x$	0	$1/2$	$\sqrt{2}/2$	$\sqrt{3}/2$	1	0	-1
$\cos x$	1	$\sqrt{3}/2$	$\sqrt{2}/2$	$1/2$	0	-1	0
$\tan x$	0	$1/\sqrt{3}$	1	$\sqrt{3}$	*	0	*
$\sec x$	1	$2\sqrt{3}$	$\sqrt{2}$	2	*	-1	*
$\csc x$	*	2	$\sqrt{2}$	$2/\sqrt{3}$	1	*	-1
$\cot x$	*	$\sqrt{3}$	1	$1/\sqrt{3}$	0	*	0

Trigonometric Identities:

$$\sin^2 x + \cos^2 x = 1$$

$$\sin(-x) = -\sin(x)$$

$$\cos(-x) = \cos(x)$$

$$1 + \tan^2 x = \sec^2 x \qquad 1 + \cot^2 x = \csc^2 x$$

$$\sin 2\theta = 2\sin\theta\cos\theta \qquad \sin(a+b) = \sin a \cos b + \cos a \sin b$$

$$\cos 2\theta = \cos^2\theta - \sin^2\theta \qquad \cos(a+b) = \cos a \cos b - \sin a \sin b$$

Lines, Circles, Ellipses, Parabolas:

* Slope of the line through (x_0, y_0) and (x_1, y_1): $m = \dfrac{y_1 - y_0}{x_1 - x_0}$

* Point-Slope Form: $y - y_0 = m(x - x_0)$

* Slope-Intercept Form: $y = mx + b$

* $(x - a)^2 + (y - b)^2 = r^2$ gives a circle, with radius r, center (a, b).

* $\dfrac{x^2}{a^2} + \dfrac{y^2}{b^2} = 1$ gives an ellipse through the points $(a, 0), (-a, 0), (b, 0)$, and $(-b, 0)$.

* $\dfrac{x^2}{a^2} - \dfrac{y^2}{b^2} = 1$ gives a hyperbola.

* $y = ax^2 + bx + c$ gives a parabola.

* Other equations involving x, y, x^2, y^2 also give one of above.

Limits:

How to find $\lim\limits_{x \to b} f(x)$

* Try to plug b into the function. If you get a number (and no 0 in the denominator or a negative number inside a square root), and if the function doesn't change its definition at b, then $f(b)$ is the limit.

* If you plug in and get $\dfrac{0}{0}$ more work must be done, usually some cancellation.

* $\lim\limits_{x \to 0} \dfrac{\sin x}{x} = 1$

Continuity:

* $f(x)$ is *continuous at* $x = a$ if
  1. $f(a)$ is defined
  2. $\lim\limits_{x \to a} f(x)$ exists and
  3. $\lim\limits_{x \to a} f(x) = f(a)$.

Definition of Derivative:

The derivative of $g(x)$ evaluated at x is written $g'(x)$ and is given by:

$$g'(x) = \lim_{h \to 0} \frac{g(x+h) - g(x)}{h} \quad \text{or}$$

$$g'(x) = \lim_{\Delta x \to 0} \frac{g(x + \Delta x) - g(x)}{\Delta x} \quad \text{or}$$

$$g'(x) = \lim_{z \to x} \frac{g(z) - g(x)}{z - x}$$

Differentiation Rules:

* Power Rule: $\dfrac{d}{dx}(x^n) = n x^{n-1}$

* Product Rule: $\dfrac{d}{dx}(fg) = f'g + fg'$

* Quotient Rule: $\dfrac{d}{dx}\left(\dfrac{f}{g}\right) = \dfrac{f'g - fg'}{g^2}$

* Trig function derivatives:

$$\frac{d}{dx}(\sin x) = \cos x, \qquad \frac{d}{dx}(\cos x) = -\sin x$$

$$\frac{d}{dx}(\tan x) = \sec^2 x, \qquad \frac{d}{dx}(\sec x) = \sec x \tan x$$

$$\frac{d}{dx}(\csc x) = -\csc x \cot x, \qquad \frac{d}{dx}(\cot x) = -\csc^2 x$$

* The Chain Rule:

$$\frac{d}{dx} f(g(x)) = f'(g(x)) g'(x)$$

or

$$\frac{dy}{dx} = \frac{dy}{du} \frac{du}{dx}$$

How to Ace Calculus: Just the Facts

Logarithmic Differentiation:

When $f(x)$ = a messy product.
1. Take ln of both sides.
2. Simplify using laws of logs.
3. Take derivatives of both sides. Get $f'(x)/f(x)$ on the left side.
4. Multiply through by $f(x)$ to get $f'(x)$.

Applications of the Derivative:

* Velocity and acceleration

  — average velocity $= \dfrac{\text{total distance}}{\text{total time}}$

  — velocity: $v(t) = s'(t)$, where $s(t)$ is the position at time t.

  — acceleration: $a(t) = v'(t) = s''(t)$.

* Graphing:

  — $f'(x) = 0$ at a maximum or minimum of $f(x)$.

  — Maxima and minima can occur where the derivative isn't defined, as well as where $f'(x) = 0$.

  — At a local maximum $f'(x)$ goes from $+$ to $-$. At a local minimum $f'(x)$ goes from $-$ to $+$.

  — At a local maximum $f''(x) \leq 0$. At a local minimum $f''(x) \geq 0$.

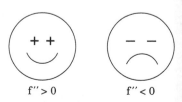

$f'' > 0 \qquad\qquad f'' < 0$

  — If $f''(x) = 0$, 2nd derivative test fails.

  — $f''(x) < 0$ means concave down (like a frown).

  — $f''(x) > 0$ means concave up (like a cup).

— At an inflection point $f''(x) = 0$.

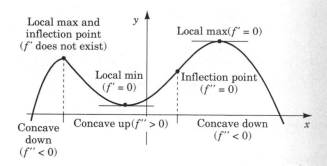

Local max and
inflection point
(f' does not exist)

Local max($f' = 0$)

Local min
($f' = 0$)

Inflection point
($f'' = 0$)

Concave
down
($f'' < 0$)

Concave up($f'' > 0$)

Concave down
($f'' < 0$)

* Differentials for approximating:

$$f(x + \Delta x) \approx f(x) + f'(x)\Delta x$$

Exps and Logs:

* $x^{a+b} = x^a \times x^b$, $(x^a)^b = x^{ab}$,
  $x^{-a} = 1/x^a$, $x^{1/2} = \sqrt{x}$

* $\ln(ab) = \ln a + \ln b$,
  $\ln(a^b) = b \ln a$, $\ln\left(\dfrac{1}{a}\right) = -\ln a$

* $b^{\log_b x} = x$, $\log_b(b^x) = x$

* $\ln e = 1$ and $\ln 1 = 0$.

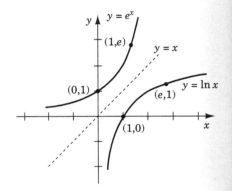

* $\dfrac{d}{dx}(e^x) = e^x$, $\displaystyle\int e^x \, dx = e^x + C$

* $\displaystyle\int e^{kx} \, dx = \dfrac{e^{kx}}{k} + C$

* $\dfrac{d}{dx}(\ln x) = \dfrac{1}{x}$, $\displaystyle\int \dfrac{1}{x} \, dx = \ln|x| + C$

* $\dfrac{d}{dx}(a^x) = a^x \ln a$, $\displaystyle\int a^x \, dx = \dfrac{a^x}{\ln a} + C$

* $\dfrac{d}{dx} \log_b x = \dfrac{1}{x \ln b}$

* Exponential growth and decay:
  $\dfrac{dN}{dt} = kN$ is the differential equation, and $N(t) = N_0 e^{kt}$ is the solution.

Integration:

* Power Rule: $\displaystyle\int x^n\,dx = \frac{x^{n+1}}{n+1} + C$ for $n \neq -1$.

* Substitution:

$$\int f'(u(x))u'(x)\,dx = f(u(x)) + C$$

* Integration by Parts:

$$\int u\,dv = uv - \int v\,du$$

* Trigonometric substitution: When expressions like $\sqrt{1 \pm x^2}$ are in the integrand, replace x by a trig function like sin or tan, simplify with trig identities.

* Partial Fractions: Rewrite a fraction of functions as a sum of fractions, and hope they are easier to integrate.

* The area under a positive function $f(x)$ between a and b is: $\int_a^b f(x)\,dx$.

* The area between $f(x)$ and $g(x)$ when $f(x) \geq g(x)$ is: $\int_a^b (f(x) - g(x))\,dx$.

* Fundamental Theorem of Calculus:

$$\int_a^b f(x)\,dx = F(b) - F(a)$$

where $F(x)$ is an antiderivative of $f(x)$, or, 2nd version:

$$\frac{d}{dx}\int_a^x f(t)\,dt = f(x)$$

* Don't forget the constant $+C$.

* Rectangle Rule:
$\displaystyle\int_a^b f(x)\,dx \approx \frac{b-a}{n}[f(x_1)+f(x_2)+\ldots+f(x_n)]$; x_i are either right-hand or left-hand endpoints of the n intervals.

* Midpoint Rule:

$$\int_a^b f(x)\,dx \approx \frac{b-a}{n}\left[f\left(\frac{x_1+x_2}{2}\right) + f\left(\frac{x_2+x_3}{2}\right) + \ldots + f\left(\frac{x_n+x_{n+1}}{2}\right)\right];$$

x_i are the endpoints of the n intervals, from far left to far right.

* Trapezoid Rule:

$$\int_a^b f(x)\,dx \approx \frac{b-a}{n}[f(x_1)/2 + f(x_2) + \ldots + f(x_n) + f(x_{n+1})/2];$$ x_i are the
endpoints of the n intervals, from far left to far right.

* Simpson's Rule:

$$\int_a^b f(x)\,dx \approx \frac{b-a}{3n}[f(x_1) + 4f(x_2) + 2f(x_3)\ldots 2f(x_{n-1}) + 4f(x_n) +$$
$f(x_{n+1})]$; x_i are the endpoints of the n intervals, from far left to
far right, and $n+1$ must be odd.

* The Definite Integral Definition:

$$\int_a^b f(x)\,dx = \lim_{n\to\infty} \sum_{i=1}^{n} f(x_i)\Delta x \quad \text{or}$$

$$\int_a^b f(x)\,dx = \lim_{n\to\infty} \sum_{i=1}^{n} f(c_i)(x_{i+1} - x_i)$$

where C_i is in $[x_i, x_{i+1}]$.

Integrals:

* $\displaystyle \int \sin u \, du = -\cos u + C$

* $\displaystyle \int \cos u \, du = \sin u + C$

* $\displaystyle \int \tan u \, du = \ln|\sec u| + C$

* $\displaystyle \int \csc u \, du = \ln|\csc u - \cot u| + C$

* $\displaystyle \int \sec u \, du = \ln|\sec u + \tan u| + C$

* $\displaystyle \int \cot u \, du = \ln|\sin u| + C$

* $\displaystyle \int \sec^2 u \, du = \tan u + C$

* $\displaystyle \int \csc^2 u \, du = -\cot u + C$

* $\displaystyle \int \sec u \tan u \, du = \sec u + C$

* $\displaystyle \int \csc u \cot u \, du = -\csc u + C$

* $\displaystyle \int \sin^2 u \, du = \frac{u}{2} - \frac{\sin(2u)}{4} + C$

* $\displaystyle \int \cos^2 u \, du = \frac{u}{2} + \frac{\sin(2u)}{4} + C$

* $\displaystyle \int \tan^2 u \, du = \tan u - u + C$

* $\displaystyle \int \cot^2 u \, du = -\cot u - u + C$

* $\displaystyle \int \frac{du}{\sqrt{a^2 - u^2}} = \arcsin \frac{u}{a} + C$

$$\ast \int \frac{du}{a^2 + u^2} = \frac{1}{a} \arctan \frac{u}{a} + C$$

$$\ast \int \frac{du}{a^2 - u^2} = \frac{1}{2a} \ln \left| \frac{u + a}{u - a} \right| + C$$

$$\ast \int \frac{du}{u \sqrt{u^2 - a^2}} = \frac{1}{a} \operatorname{arcsec} \left| \frac{u}{a} \right| + C$$

$$\ast \int \frac{du}{\sqrt{u^2 \pm a^2}} = \ln |u + \sqrt{u^2 \pm a^2}| + C$$

$$\ast \int \frac{du}{(u^2 \pm a^2)^{3/2}} = \pm \frac{u}{a^2 \sqrt{u^2 \pm a^2}} + C$$

$$\ast \int \ln u \, du = u \ln u - u + C$$

$$\ast \int u e^u \, du = (u - 1)e^u + C$$

$$\ast \int u \sin u \, du = \sin u - u \cos u + C$$

$$\ast \int u \cos u \, du = \cos u + u \sin u + C$$

$$\ast \int \frac{1}{u \ln u} \, du = \ln |\ln u| + C$$